Walter Alden

**The Human Eye**

Its use and abuse, a popular treatise on far, near and impaired sight, and the

methods of preservation by the proper use of spectacles and other acknowledged

aids of vision

Walter Alden

**The Human Eye**
*Its use and abuse, a popular treatise on far, near and impaired sight, and the methods of preservation by the proper use of spectacles and other acknowledged aids of vision*

ISBN/EAN: 9783337248451

Printed in Europe, USA, Canada, Australia, Japan

Cover: Foto ©berggeist007 / pixelio.de

More available books at **www.hansebooks.com**

# THE HUMAN EYE;

## ITS USE AND ABUSE:

A

POPULAR TREATISE ON FAR, NEAR AND IMPAIRED
SIGHT, AND THE METHODS OF PRESERVATION
BY THE PROPER USE OF SPECTACLES,
AND OTHER ACKNOWLEDGED
AIDS OF VISION.

## BY WALTER ALDEN,
OPTICIAN.

"The Human Eye is the most important organ of sense, at once challenging our
wonder and respect."

CINCINNATI:
PUBLISHED BY THE AUTHOR.
1866.

R. N. CARTER,
BOOK AND JOB PRINTER,
66 West Third Street, between Walnut and Vine Streets,
CINCINNATI, OHIO.

# PREFACE.

THE favorable reception accorded to, and evident interest manifested in my articles in the "*News and Educator*," upon the subject of "The Eye and Vision," and the lack of due information upon this important subject, suggested the idea of this simple treatise on "THE PRESERVATION OF THE SIGHT," "The Construction of the Eye, Near, Far, and Over Sight." This work is presented for the acceptance and approval of the public, touching a vital subject upon which little attention has hitherto been paid. The methods of preserving vision, the most precious and valuable of the senses, has been a subject too long neglected. Thousands grope along in almost total blindness, who, by proper treatment, might have preserved their sight. It has been my aim to meet the necessity for correct information upon this interesting and important subject. Avoiding, as far as possible, technicalities and a scientific nomenclature, I have endeavored to present the subject in a popular form—at once, simple, concise and interesting.

WALTER ALDEN.

CINCINNATI. 1866.

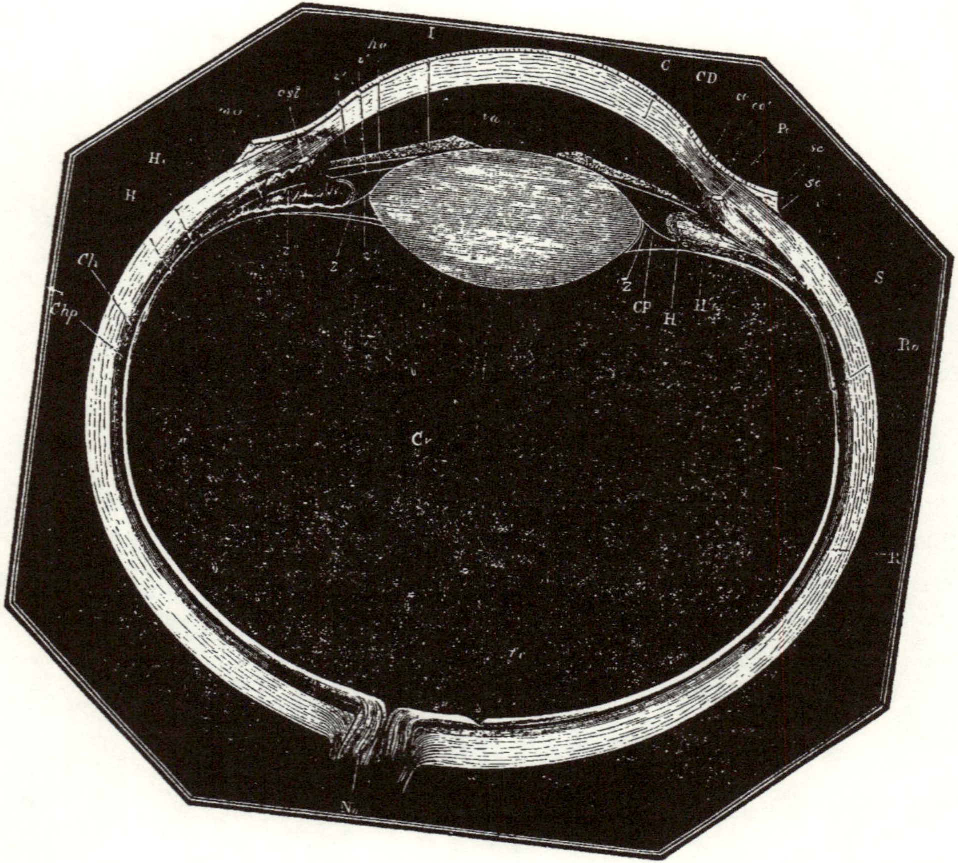
I
C
CD
S
Ro
Cv
CP
H
Z
Ch
Chp
H
osl
va

# DESCRIPTION OF FIG. 1.

S.   SCLEROTICA.
Sc.  Conjunctiva.
Sc'.  Epithelium of the Conjunctiva.
C.   CORNEA.
cc.  Anterior elastic laminæ of the cornea passing over from the layer of the connective tissue of sclerotic conjunctiva.
cc'.  Epithelium of the Cornea.
CD.  Membrane of Descemet.
Z.   Canal of Schlemm, (circular sinus.)
Ch.  CHOROID.
Chp.  Pigment layer.
Pc.  Ciliary Processes.
mci.  CILIARY MUSCLES.
I.   IRIS.
R.   RETINA.
Ro.  Ora serrata.
No.  OPTIC NERVE.
fc.  Fovea Centralis, (Yellow spot.)
H.   Hyaloid.
CP.  Canal of Petit.
L.   CRYSTALLINE LENS.
cv.  VITREOUS HUMOUR.
va.  ANTERIOR CHAMBER.
ha.  POSTERIOR CHAMBER.

# DESCRIPTION OF FIG. OPPOSITE I.

From Soelberg Wells' Impaired Vision.

## SECTION OF THE HUMAN EYE CONCERNED IN ACCOMMODATION. × 15.

••

C.      Cornea.

ce      Its anterior elastic laminæ passing over from the layer of connective tissue (sc) of the sclerotic.

cc'.    Epithelium of the anterior surface of the Cornea.

CD.     Membrane of Descemet; its homogeneous lamella.

CDE.    Its epithelium.

S.      Sclerotic.

sc.     Vascular layer of connective tissue of the sclerotic conjunctiva.

sc.     Its epithelium.

csl.    Canal of Schlemm.

Ch.     Choroid.

Cp.     Its pigment layer.

Pc.     Ciliary Process (a longitudinal section).

mci.    Ciliary Muscle.

I.      Iris.

ip.     Pigment layer.

R.      Retina.

Ro.     Ora serrata.

H.      Hyaloid.

H.*     Place of its division.

H'.     Zonula.

Z.      Its free portion.

H".     Posterior laminæ of the hyaloid.

cp.     Canal of Petit.

L.      Lens.

Lc.     Capsule of lens.

va.     Anterior Chamber.

ha.     Posterior Chamber.

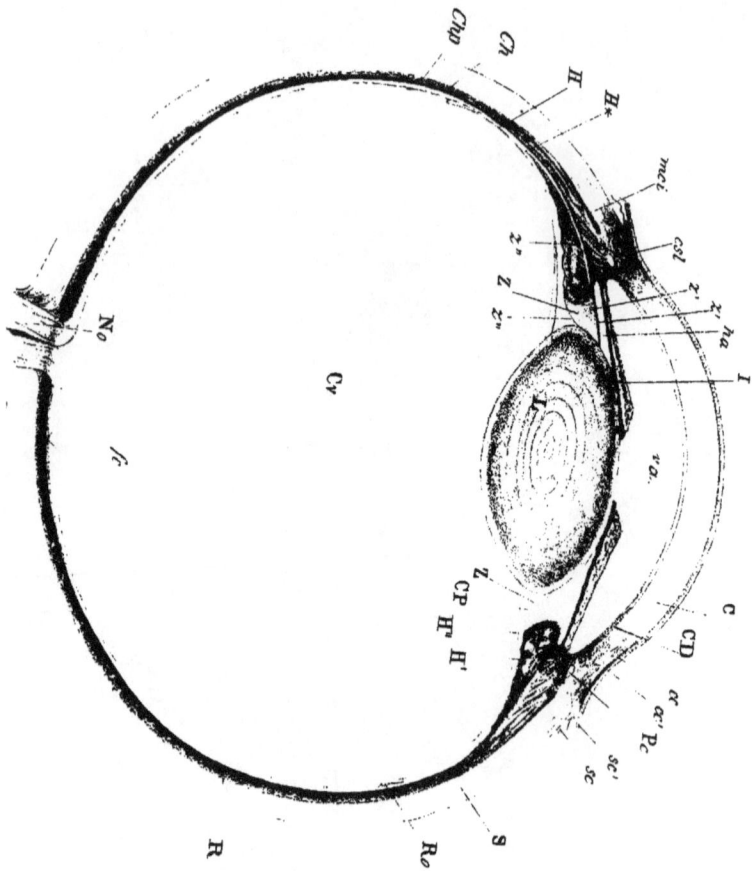

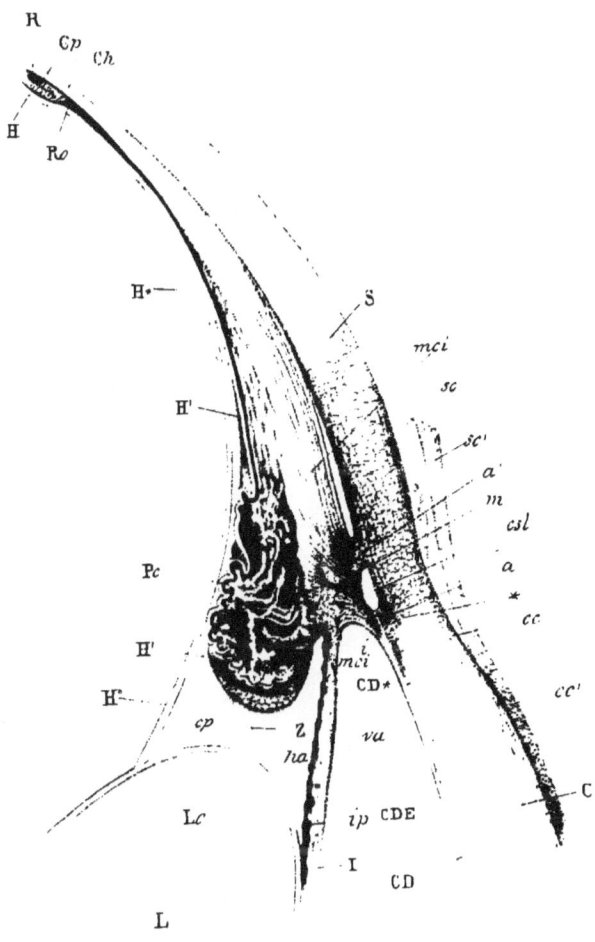

# THE HUMAN EYE;

## ITS USE AND ABUSE.

## CHAPTER I.

### DISSECTION OF THE EYE.

But few persons ever spare the time to consider or
observe the human eye, or to acquaint themselves with
the laws that govern it, and the best method of preserving its functions unimpaired for the greatest length of
time. And it is to this indifference, or rather neglect,
that so many individuals are indebted for their impaired
vision or blindness. No man, in his right mind, would
allow his arm or any other member of his body to waste
away, because kept continually on the strain, and yet
thousands become partially blind, merely because the
eye is thoughtlessly permitted to labor and exhaust its
powers of endurance. The human eye is one of the
most important organs of sense, at once challenging our
wonder and admiration when we observe its beauty, the
excellence of its structure, and the perfection with which
it performs all its allotted functions. In its construction, it is a true camera obscura, through which the mind
is placed in direct communication with the external
world, and is enabled to observe and investigate *motion,
magnitude* and *positions*. Without its assistance the mate-

rial world would be a sealed book to us, of which we
could have but little, if any, conception.

When the eyeball is removed from the socket and
investigated, we find it to be a hollow sphere enclosing
certain fluid and semi-solid substances. The wall or shell
of this sphere, though not very thick, is composed of
several layers or coats, each one having its own functions
to perform in the preservation and maintenance of the
organ and its functions.

The outermost coat (see Plate 1, for general positions
of parts here described) of all is called the *sclerotica*, or
*sclerotic* (S) *coat*, which is a tough, white, resisting membrane, serving to sustain the delicate parts within; it
composes that part of the front surface of the eye, known
as the *white of the eye*, and is covered externally with a
transparent, very fine mucous membrane, called the *conjunctiva* (Sc), and which membrane is usually the seat of
disease in the more ordinary inflammations of the eye.
The sclerotica encloses the whole of the eyeball, with the
exception of that part in front, which bulges out like a
watch glass, and appears as if framed into the sclerotica;
this part is called the cornea (c), from its tough, hard,
and horn-like character, being derived from the Latin,
*cornua*, signifying horn. The cornea is about 9-20ths of
an inch in diameter, is also covered externally with a
very thin and delicate layer of the conjunctiva, is more
convex than the rest of the eye, and, from its clearness
and transparency, permits the light to pass through into
the interior parts of the organ, hence it has been not
inaptly styled "the window of the eye."

The second coat of the eye, is that which lines the
inner surface of the sclerotica, and is called the *choroid*,
(ch). It is a still more delicate membrane, consisting
chiefly of blood vessels, and cells which form a slimy,

granular, intensely black substance upon the inner surface of the choroid coat, and which is called "black pigment," or *pigmentum nigrum*, (chp). This pigment gives
to the eye a jet black substance, which absorbs the extra-
neous rays of light, prevents reflection, and thus aids in
perceiving the distinctness of the image formed on the
retina. In Albinos there is a deficiency of this pigment,
in consequence of which vision is indistinct with them,
and in a bright light they experience a painful dazzling.

The *retina* (R) is a very thin, transparent expansion
of the optic nerve, forming the innermost coat or layer
of the back part of the eyeball. Prof. YOUMANS
observes: At the back part of the *eye*, the sclerotic coat
is formed into a tube which leads inward to the brain.
This tube contains the *optic nerve*. As it enters the
globe, it spreads out over the inner surface of the choroid,
in the form of a most delicate network of nervous fila-
ments, called, from its reticulated structure, the *retina*.
The retina is therefore the extended and diffused optic
nerve. In dissection it is easily separated from the
choroid. It is absolutely transparent so that light and
colors penetrate and *pass through* it perfectly, and therefore fall upon the dark surface beneath. To prevent the
delicate and transparent tissues of the retina from being
stained by the black pigment, a very thin film is inter-
posed between them, called "Jacob's membrane." As
delicate and as thin as is the retina, it is composed of five
distinct layers; and, about one-tenth of an inch, externally, from the entrance of the optic nerve is a transparent, circular, and yellow spot. At the place where the
optic nerve enters the eye, the nerve fibres are totally
insensible to light; and the images transmitted to the
brain first pass through the retina and form on the black
pigment.

Having now briefly examined the layers of the wall of the eyeball, let us devote a short time to a consideration of the contents of its cavity.

Immediately behind the cornea is a space filled with a perfectly clear and colorless liquid called the *aqueous humor*, (va), which is contained in a cavity that is convex in front and concave at its back part, and which extends from the inner concave surface of the cornea to the outer convex surface of the crystalline lens. This cavity or space is divided into two smaller cavities by the iris, and which communicate freely with each other through the pupil; the front one is called the *anterior chamber of the aqueous humor*, and the back one, the *posterior chamber*.

The *iris* (I) is a thin disk or circular membrane, which forms the partition between the two chambers just referred to; its central part is perforated by an aperture, called the *pupil*, which is simply a hole through the iris. The iris is colored differently in different persons, and according to its color, we designate the eyes as black, blue, gray, hazel, etc. It likewise possesses the remarkable property of dilating and contracting, so as to admit the necessary amount of light into the eye for the function of vision, independent of the will; in the dark the iris expands, thus admitting more light to enter, and thereby enabling us to distinguish objects which, but for this beautiful mechanism, we would be unable to see— its expansions forms the *dilated pupil*. In the full light it contracts, permitting only the necessary rays of light to enter the eye for vision, and then forms the *contracted pupil*. This can be readily demonstrated by any one holding a mirror before his eyes, and then moving a candle to and from them; as the candle is removed from the eyes, the iris will be observed to dilate, as it advances toward them, this membrane will be seen to contract.

At the back part of the posterior chamber of the aqueous humor, is a hard, jelly-like, perfectly transparent substance, called the *crystalline lens*, (L); it is a double convex body, consisting of many concentric layers, resembling somewhat those of an onion, having its front surface less convex than its back surface, and is suspended in a transparent bag or capsule by a ring of muscles placed around it, called the ciliary processes (mci). The use of the crystalline lens is to refract the rays of light, and bring them to a focus upon the retina; also to correct spherical aberration "by the varying density of the lens, which, having a greater refractive power near its center, refracts the central rays in each pencil of light to the same point as its circumferential rays.— *Sir D. Brewster.*

The space behind the crystalline lens, constitutes the greater portion of the cavity of the eyeball, and is filled with very delicately organized cells, containing a clear, transparent, soft, jelly-like mass, which is concave in front and convex behind, and is termed the *vitreous humor*. This is of less refractive power than the crystalline lens, and gives passage to the converging rays of light, slightly increasing their convergency; it also "excludes the heat rays, so as to preserve, under all circumstances of exposure, a uniform condition and effect upon the retina."

The freedom from color in the images formed on the retina, is due to the achromatic combination of the lens and humors of the eye. The aqueous and vitreous humors form two menisci of different densities, and the double convex crystalline lens placed between them is of greater density than either, thus forming a most perfect achromatic combination.

# CHAPTER II.

### ACCOMMODATION OF THE EYE.

The peculiar, unconscious action of the healthy eye, by which it becomes so adjusted as to see objects at various distances from it in the field of vision, is termed the "power of accommodation" or adjustment of the eye, or simply "accommodation." That the eye possesses this power of adjustment, acting almost instantaneously and without apparent effort, is self-evident to every one, and requires no argument nor experiment to demonstrate it. But how this adjustment occurs, or what changes occur in the eye to effect it, has long been a matter of excited debate among physiologists. However, it has more recently been proven, by close observation and research, that the condition of those parts of the eye essential to vision differ materially in their positions and relations to each other, according as objects looked at are near the eye, or at a distance from it. It is well known that, on looking through a microscope or telescope, or even a common lens, we have to adjust or accommodate the instrument "to suit our eye," that we may distinctly see objects at different distances in the field of vision, or within the range of accommodation, and, the same as with the eye, the position and relation of the dioptric parts of the instrument must vary according to the distances of the objects looked at.

It is now settled beyond a doubt, that this power of accommodation of the eye is due to a change in the form or convexity of the crystalline lens ; so that, in other words, the accommodation of the eye is simply the power of increasing the convexity of its crystalline lens. When the eye is in a state of rest, it is accommodated or adjusted for objects at a distance ; but when these objects are brought nearer to the eye, the crystalline lens involuntarily increases in its convexity, more especially upon its anterior surface, while at the same time it approaches somewhat nearer to the cornea.

That the dioptric apparatus of the eye really undergo a change in their relations with each other, when the condition of adjustment for seeing near and distant objects is effected, is well illustrated by the following experiment from Donders: Hold a veil at some inches from the eye, and a book at a greater distance; we can then at will see accurately either the texture of the veil, or the letters of the book, but never both together. If we see the texture of the veil, we cannot distinguish the letters of the book; if we read, the veil produces only a feeble, almost uniform obscuration of the field of vision, but we see scarcely anything of its separate threads. The circle of diffusion* in imperfect accommodation can be most distinctly seen at an illuminated point, or at a darker spot on a piece of ordinary window glass. The latter is held close to one eye, (while the other is shut,)

---

* When each point of an object looked at is no longer represented upon the retina by a point, but by a CIRCLE (the section of each conical pencil of rays), these circles from contiguous points necessarily overlay, and the object looked at, instead of appearing well defined and distinct, becomes blurred and indistinct. These circles are termed "circles of diffusion." When we look at near objects the pupil contracts involuntarily, and when we look at distant ones it dilates, and, therefore, it may be that the iris acts the part of a self-changing diaphragm, diminishing any slight circles of diffusion in the image on the retina, and thus becoming a corrective supplement to our accommodation.

but so that the point can still be accurately perceived— the objects situated at a certain distance on the other side of the glass are then seen without defined contours or outlines, they appear blurred and indistinct. We can now, however, at will, immediately see, in the direction of the point, the objects at the remote side of the glass distinctly, whereupon the point appears as a larger, diffused spot. A change has consequently taken place in the eye, of which we are ourself distinctly conscious. When we looked at distant objects through the glass, our eye was adjusted for almost parallel rays, or for objects at a distance; the diverging rays proceeding from the point, had therefore their point of union behind the retina. When the point was accurately seen, the eye was accommodated to the diverging rays proceeding from it, and the almost parallel rays derived from the distant objects had already united in front of the retina and had decussated in a focus. In uniting, whether before or behind the retina, the rays proceeding from each separate point formed a *round* spot on the retina, instead of a point. The section of these rays has, in fact, nearly the *form of the pupil*, and if the rays of the cone have not yet been brought into union, or if they have already decussated, they form on the retina a little spot of the form of the pupil. All the little spots, which represent the several points of the object in the retinal picture, are now like so many blotted points of an accurate image covering one another,* and it is evident that the former must, therefore, lose its sharp contour and be diffused on the surface. But, as the retinal, so is the projected picture, and we therefore say, that we see the object diffused or blurred. And all objects, for which

--------

* Forming circles of diffusion.

the eye is not accommodated, appear in such a condition.

Fig. 2.

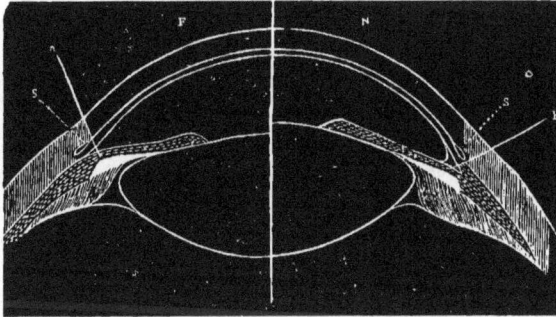

The changes undergone by the dioptric apparatus of the eye during accommodation, are well illustrated in the accompanying Fig. 2. The front part of the eye is divided into two equal parts by the vertical line between them. The one-half, F, shows the position of the parts and especially of the crystalline lens when the eye is adjusted for distant objects; the other half, N, when it is adjusted for near objects. When the eye is in a state of rest, as in the half, F, the iris forms a curve, as at (a) in the vicinity of Schlemm's canal at (s); but when the accommodation is for near objects, as in the half, N, the fibres of the iris become contracted, the periphery of the iris straightened (b), and the anterior chamber lengthened, thus making up for its loss in depth, through the advance or protrusion forwards of the front surface of the lens.

The means by which the change in the form of the crystalline lens is effected, is a question still discussed by physiologists, and one that has greatly occupied the

attention of the most eminent oculists of Europe and this
country for a number of years. It is now, however,
more generally considered, (by DONDERS, GRAEFE, POR-
TERFIELD, J. S. WELLS, J. Z. LAWRENCE, HEMHOLTZ,
WECKER, etc., etc.,) that this change in the form of the
lens is effected by the action of the ciliary muscles.

*Range of Accommodation* means the distance between
the nearest and farthest points at which objects may be
distinctly seen by a person. In a healthy eye the nearest point of distinct vision averages about 3 or 4 inches
from the eye, while its far-point, termed by oculists
"infinite distance," may vary. Any change or derange-
ment of the accommodative power of the eye, of its
range of accommodation, or of its refraction or refractive
power, gives rise to certain abnormal conditions of
vision, some of which I propose to briefly consider.

*Refraction* means the deviation of the rays of light
from a straight line, and although intimately associated
with the conditions of vision, yet it is wholly different
from "accommodation." The different refractive media
of the eye are, commencing from without, the cornea,
aqueous humor, crystalline lens, and vitreous humor;
and the form and relations of these with each other
effect the refractive power of the eye by bringing parallel rays in front of, upon, or behind the retina, producing myopia, emmetropia or normal vision, or hypermetropia. Presbyopia, or the long sightedness of age,
is not due to a morbid condition of refraction, but to a
diminution in the range of accommodation, probably
from impairment of the contractile energy of the ciliary
muscle, which renders it incapable of producing sufficient convexity of the crystalline lens for distinguishing
very near objects. Refraction is normal, when the eye,
in a state of rest, or adjusted for its far point, brings par-

allel incident rays to a focus upon the retina, without any active effort of the accommodative power. The following table exhibits a system of classification founded upon the refractive condition of the eye when in a state of rest, [J. Z. LAWRENCE,] and which I design discussing hereafter:

*The eye in a state of rest, in which the crystalline lens is at its minimum curvature, and the optic axes are parallel.*

| 1. | 2. | 3. | 4. | 5. |
|---|---|---|---|---|
| EYE. | Parallel rays are focussed | Far point : | Eye in a state of rest adapted for : | Effect of glasses for distant objects : |
| 1. Normal or Emmetropic | On the retina. | At an infinite distance | Parallel rays. | Convexes & concaves deteriorate vision. |
| 2. Myopic. | In front of the retina. | At a definite distance and positive. | Divergent rays. | Concaves improve vision. |
| 3. Hypermetropic. | Behind the retina. | At a definite distance and negative. | Convergent rays. | Convexes improve vision. |

The three lowermost lines in the above table, read horizontally, give the leading characteristics of each class of eye; read vertically, the leading contrasts.— J. Z. LAWRENCE.

# CHAPTER III.

FREQUENT CAUSES OF IMPAIRED VISION.

THE eye, as we have seen from the preceding chapters, is a delicate structure, exceedingly sensitive, yet performing an immense amount of labor without any appreciable fatigue, and adapting itself to various distances, and to various intensities of light, so as to astonish even the most casual observer. Thus, we can read by the light of the moon, by twilight, by the light from a small taper, &c., and these lights are by no means to be compared with the light emanating from the sun. The light from the sun is estimated to be 300,000 times more brilliant than that from the moon; or equal to that given out by 5,000 wax candles of moderate size, supposed to be placed at the distance of one foot from the object looked at. Moonlight is supposed to be about equal to that from one candle at the distance of twelve feet.

The eyeballs are securely enclosed within a bony socket, and are so arranged, that by means of their muscles, six in number to each eye, they may move in concert and be directed to the several objects in the field of view. They are composed of walls, vessels, membranes, muscles and nerves, and are nourished by the vital blood-stream, the same as any other organ ; and as with other organs, when they are in active use, they expend force, exhaust themselves, and become fatigued. But, unlike any other organ of the body, their powers of endurance, of undergoing active exercise, is remarkably great; indeed, were it not for this power, but few would reach the prime of life with unimpaired vision. And even as it is, when we consider the Herculean labor that the eye is called upon to perform, and the numerous abuses to which it is subjected, the wonder is, not that there are so many weak, inflamed and sightless eyes, but that there is not a multitude more of them.

The eyes are injured by a great variety of causes, and when constantly exposed to these causes, vision becomes more or less permanently impaired. Intense light, artificial light, flickering light, are all detrimental to vision; the best, most agreeable, and least injurious to the eyes, is daylight. I will now notice a few of the more common causes producing injury to the sight. After a severe illness, while the eye is yet recovering from the debility impressed upon it by the constitutional suffering, and is unable to bear fatigue, it is frequently the case that convalescents, with whom time hangs heavily, endeavor to occupy themselves, and employ their time in reading—reading even more constantly than when in health. This is very wrong; it inflicts a serious injury upon the nervous power of the eye, and from this over-exertion, a partial blindness frequently results, which is often permanent and beyond the power of the oculist to cure.

While occupied in reading some interesting book, twilight comes upon us; instead of ceasing and placing the book aside until a better light can be had, how often is it the case, that individuals continue to read, and, as the twilight deepens, strain the eye to the utmost in order to see page after page, until finally the letters become confused, mingle with each other, and a mist gathers before the eyes! This is a common but very pernicious habit.

Travelers in railroad cars often supply themselves with reading matter in order to pass away the time; but the constant motion and jolting of the cars, render the letters more or less movable, as it were, and indistinct, requiring an immense straining of the eyes to make them out, so that but few persons can read in this manner, to exceed an hour at a time, the eye becoming fatigued, hyperemied, and the lids becoming heavy, with a greater or less sense of drowsiness. This is an exceedingly injurious course, and those who wish to preserve sight to an advanced age had better suffer ennui than to thus strain and injure vision. For a somewhat similar reason, persons should never read by a flickering light, nor under a tree, where, from the motion of the leaves by the wind, lights and shadows are constantly and irregularly encountered; nor while walking, riding in a carriage, &c. In these instances, the continual change of, and strain upon, the accommodative function of the eye, in conjunction with the overtaxing of the retinal function, are the principal causes of the injury to the organ. Daylight of itself, never wavers or flickers, its action upon the eye is unvarying, and beneficial. Under the influence of daylight, the pupil and the retina adapt themselves to the uniform flow of light, and the eye undergoes no fatigue from proper use. On the other hand, sudden variations of light do not allow the pupil and the retina time to adapt themselves to these sudden changes, and the eye becomes strained and injured, as may be known by the impaired distinctness of vision, the temporary blindness, and other unpleasant symptoms resulting therefrom.

Reading, or working in the brightness of sunshine, or with a bright, dazzling artificial light, dazzles and confuses vision, pains the eyes, and frequently exerts a permanently injurious effect upon the retina, causing partial or complete blindness. The same may be said, where a light is held between the eye and paper to enable the party to read or work to better advantage. In all cases, for the preservation of sight, the source of light should invariably be placed behind the eyes, or to one side of them, but never in front of them.

One of the most frequent causes of blindness is "inflammation of the optic nervous apparatus," the result of exposures of the eyes to intense lights, and consequent overstraining of the sight. Persons have been struck blind while looking at an eclipse, without properly protecting the eye; others from the reflection of light from snow, and hence the general use of colored spectacles by the Canadians. I will state here that cobalt blue glasses are the best and most proper to prevent the dazzlings caused by too intense a light, as they exclude the most irritating rays of the spectrum, without plunging one into a state of semi-obscurity, and which is the fault of the gray glasses, and London smokes, which eliminate indifferently a variable quantity of all the rays of the solar spectrum.

Continued reading or examinations of objects by dim or flickering lights; or after a sleepless night; or after an illness, are fruitful causes of injury to the eye. Dr. ELLIOTT states that nearly two-thirds of all the cases of this disease (amaurosis) that he met with in practice, occurred "among those who use their eyes a great deal by artificial light, such as literary men, students, compositors, tailors, seamstresses, shoemakers, engravers, glass-blowers, &c., &c." And whenever a person finds his sight failing from an exposure to any of the above named causes, he should at once apply to a scientific oculist, for procrastination or delay is dangerous.

The following cases are given in " MACKENZIE on the Eye," page 982:

CASE I.

" The celebrated Dr. REID, Professor of Moral Philosophy, in the University of Glasgow, in May, 1761, being occupied in making an exact meridian, in order to observe the transit of Venus, rashly directed to the sun, by his right eye, the cross-hairs of a small telescope. He had often done the like in his younger days with impunity, but suffered from it on this occasion. He soon observed a remarkable dimness in that eye; and for many weeks, when he was in the dark, or shut his eyes, there appeared before the right eye a lucid spot, which trembled like the image of the sun seen by reflection from water. This appearance grew fainter and less frequent by degrees, but some very sensible effect of the injury remained. The sight continued dim; the nearest limit of distinct vision was rendered more remote than in the other eye; and a straight line appeared to the right eye to have a curvature in it."

CASE II.

Again, "Mr. ALLEN mentions the case of a master of a printing office, who became blind. He had corrected the press, and was otherwise engaged in reading, for 18 hours out of the 24, a practice which he continued for twelve months, notwithstanding an evident failure of his sight. At the end of this time, the amaurosis was so complete that he could not distinguish one object from another, but was merely capable of perceiving the light so as to find his way into the streets. He continued in this state for several years, but ultimately recovered sight."

The use of tobacco in any form, but more especially in smoking, affects the eye seriously. If we try to write or read, and smoke at the same time, we find the effort to be exceedingly painful. If we let tobacco alone, the strength of the eye will be increased, and its power preserved intact for many years. Tobacco gradually, little by little, undermines the sensitiveness of this delicate organ, irritating its conjunctival lining membrane, and paralyzing its nervous power, besides exerting a similar influence upon the brain, and nervous system of the body.

Pressure on the eyeball, or improper rubbing by the fingers, so frequently done by thoughtless persons, often occasions permanent injury to the eye. BEER relates the following:

CASE III.

"A man who previously enjoyed excellent sight, happened to be in company with some friends, when suddenly a stranger slipped behind him, and clapped his hands upon his eyes, desiring him to tell who stood behind him. Unable or unwilling to answer the question, he endeavored to remove the hand of the other person, who only pressed them the firmer on the eyes, till at length withdrawing his hands so as to allow the eyes to be opened, the man found that he saw nothing, and continued ever afterwards blind, without any (apparent) lesion of sight."

Another, and one of the principal abuses to which the

eyes are subjected, is the careless and improper use of spectacles. Many persons, in fact the majority of community, as soon as they find the sight commencing to fail, instead of applying to a skillful and experienced oculist, or optician, that they may be furnished with the proper glasses to improve and preserve vision, thoughtlessly purchase a pair of spectacles from a pedlar, or from a jeweler, who is entirely ignorant of the first principles of Optics, as well as of the proper requirements of the eye, erroneously imagining that, no matter how much the glasses may magnify, if they can only read, sew or work with them, that is all that is required—little dreaming how great and irreparable an injury is done to the eyes by the use of unsuitable glasses. A case has come under my own observation, where a woman sent her little girl to buy a pair of spectacles!•

Generally, persons ignorant of what is required, adapt the *eye to the spectacles*, instead of adapting the *spectacles to the peculiar condition of the eye*. It is a prevalent, but erroneous and injurious supposition that when the sight fails, *magnifying* spectacles are needed. The rule should always be to select those glasses which enable the person to distinctly read the finest print at about ten inches from the eye, but which *do not magnify at all*. We should ever bear in mind that all that is required in the use of spectacles is *distinctness of vision without regard to size*.

The celebrated English oculist, J. SOELBERG WELLS, makes the following very pertinent remarks upon this subject:—" The proper and scientific choice of spectacles is, indeed, of great importance to the public; and I have no hesitation in saying that the empirical, haphazard plan of selection generally employed by those who sell spectacles is but too frequently attended by the worst consequences, that eyes are often ruined which might, by scientific and skillful treatment, have been preserved for years. I would, therefore, strongly recommend the adoption of the following plan, which is largely employed on the Continent, and also by several ophthalmologists in England:—The medical man (oculist) himself, selects the proper glass from his spectacle box [which contains concave and convex glasses, corresponding numbers being kept by the optician]; the focal distance of the required glass [for one or both eyes] is written on a slip of paper, which is taken to the optician, who supplies the patient with the spectacles prescribed thereon. Thus are we sure that the person is furnished with the proper glasses." It is just as improper to wear glasses too weak, as too strong: both straining and injuring the eye.

The eye is often abused by an exactly contrary course. Many persons erroneously believing it right to do without glasses as long as possible, allow their eyes to become strained and exhausted to a great extent, and will not wear spectacles until they are actually compelled to. This course is wrong, as amaurosis, diseases of the retina, cataract, blindness, &c., are frequently occasioned by it. Whenever a person cannot read at his usual distance without the eyes becoming fatigued or painful, and the letters becoming indistinct or blurred, he should at once procure the proper spectacles suited to the condition of his eyes, and by this course he will preserve his vision for a much longer period. If he, on the contrary, allows the failure of sight to go on without making use of correct means to remedy it, he must expect not only an increased impairment of vision, but in many cases, an attack of some one of the diseases already mentioned.

And now having thus briefly examined the many causes of the impairment and weakness of vision, let us turn to the consideration of natural defects and their proper treatment.

# CHAPTER IV.

## MYOPIA OR NEAR-SIGHTEDNESS.

In READING ordinary-sized type, the correct distance of vision is 10 or 12 inches; when the book has to be held nearer the eye than this, in order to enable the person to read distinctly, he is said to be near-sighted or *myopic;* if it has to be held farther off, he is then said to be far-sighted or *presbyopic.*

In cases of myopia, (near-sightedness) the rays of light, which may be easily seen in

FIG. 3.

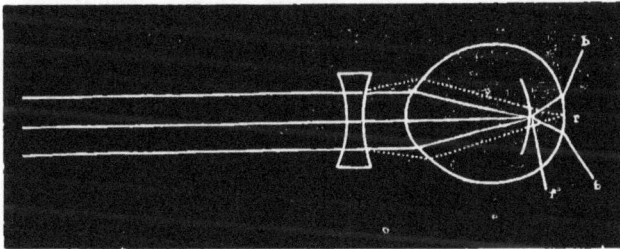

are too powerfully refracted, so that instead of being brought to a focus upon the retina, they do not reach this membrane, but fall short of it, in consequence of which the image of an object looked at is formed in front of the retina, and appears confused, indistinct, and larger than natural; and, in order to see it distinctly, the object has to be placed nearer the eye. The rays of light from the object are now more divergent than when it was at a distance, and do not come to a focus so soon, but are thrown back upon the retina, so that the image is seen clearly and distinctly. A similar result occurs in cases where the optic axis is longer than usual, or, where there is too great a distance between the lens and the retina.

In myopic persons we frequently observe considerable convexity of the cornea, and the eyes are more prominent than the healthy eyes, the pupil extended, the eye-balls firm, and the eye-lids tender; but this is by no means invariable, as many myopic eyes present no unusual appearances. But, it must be borne in mind, that myopia is often an indication of a disease of the inner eye, termed "sclerotico-choroiditis posterior," and which is due to a thinning of the back part of the choroid and sclerotica, and a consequent protrusion posteriorly of a part of the walls of the eye in proximity with the optic nerve. In this case, the myopia advances with the progress of the disease, until the patient ultimately becomes blind; or, should the disease remain stationary, the myopia will also. This is a very serious affection of the eye, and can only be detected by the ophthalmoscope, an instrument designed for seeing the internal parts of the eye; hence, the great advantage of a myopic person consulting a good oculist as early as possible, in order to learn whether his myopia be due to this disease or not; that he may at once be placed under appropriate treatment to save the eyes, if this malady be present. For although spectacles may for a time relieve vision, yet they will require to be changed often, and at short inter-

vals, until vision fails entirely; other means are, therefore, necessary in myopia from this cause.

It has been noticed that myopic persons usually see
better in a weak and dim light, that they are apt to
write a small cramped hand, and they prefer reading
small type, "the enlargement of the visual angles by
the proximity of the object enabling them to do so with
ease," or, in other words, the objects being nearer the
eye, this organ receives a greater quantity of rays from
them.

It has also been remarked that they do not look a
person in the face, when in conversation, because the
motions of the eyes and mouth of the person in speaking
are not seen by them. A strong light enables a myope
to see at a greater distance than a weak light, on account
of the contraction of the pupil occasioned, thus excluding
all but the more direct rays. On this same principle
they partially close the lids on attempting to look at distant objects. Upon looking through a pin-hole in a card,
they will also observe distant objects more distinctly
than without the aid of the pin-hole.

The myope (short-sighted person) sees near objects
much better than the normal eye, so that near-sightedness becomes, in a measure, rather an advantage than a
detriment, especially for those who are required to work
on fine structures, minute objects, etc., as they can
observe all the details of their work very accurately;
and when it is required for them to look at distant
objects, they can readily obtain the proper assistance by
means of appropriate spectacles.—Many persons, with,
originally, healthy eyes, become myopic, through continued and constant exercise of these organs on fine
work; hence, we find myopia more prevalent in the
higher than the humbler walks of life. Laborers,

mechanics, soldiers, sailors, etc., are rarely myopic; while students and professional men are very apt to be. This statement is corroborated by Mr. WARE, who writes that upon careful inquiry, he found "that out of three regiments of foot-guards, consisting of nearly ten thousand men, in the space of twenty years not six individuals had been discharged on account of this imperfection, and, at the military School at Chelsea, among 1,300 children, near-sightedness had never been known." Whereas, "in one college at Oxford, where the society consisted of 127 members, 32 either wore spectacles or used hand-glasses ! ! "

School teachers, who have been confined to their school rooms for a number of hours each day, have often found their vision limited to the extent of the walls of these rooms, the eyes having lost the power of accommodation for a greater distance. The eye obeys the same laws as the rest of the body—the proper use of any organ develops it and sustains its functions; while a want of use, or an improper use of it, enfeebles it, causes it to waste, or impairs it in some other way. Muscles are developed by constant and proper use, as observed in the arm of the blacksmith, the leg of the fencer, etc.; so is the function of vision benefitted by a proper use of the eye.

Short-sightedness is usually hereditary, but is rarely annoying until at the age of puberty, when it may continue to increase. If, when it is first noticed, prudence and perseverance were exercised, to accustom the eyes to accommodation for long distances, avoiding fine work, reading fine print, etc., the myopia might be partially, if not completely, cured. But, instead of this, children are made to pore over their books, hour after hour, seldom having permission or opportunity to look at distant objects, or to exercise the eyes on scenes in the open country—thus unwittingly increasing, instead of diminishing the near-sightedness.

Myopes generally attribute to distant objects a greater magnitude than really exists; this arises from the fact that in the healthy eye, the images are formed at the intersection of the refracted rays of light issuing from the object, while in the myopic eye, the retina receives the image of the object at a short distance behind the intersection of these refracted rays, in consequence of which, they are rendered not only dim and indistinct, but also more extended. Near-sighted persons frequently state that although their sight is excellent for near, it fails when applied to distant objects, they often nip their eye-lids close together, the reason for which is two-fold : (1) they, by this means, narrow the opening between the lids, and thus cut off some of the peripheral rays of light, and consequently diminish the circles of dispersion on the retina; (2) they thus exercise a certain amount of pressure upon the eye-ball, the cornea becomes somewhat flatter, and the far point removed further from the eye, the latter, therefore, becoming less myopic."—*J. Sœlberg Wells.*

The common error into which people fall, in regard to myopia, is, of supposing that, as the myope advances in years, his short-sightedness will, gradually, diminish, and vision become improved. Such beneficial changes in the eye, very rarely occur; on the contrary, many instances have come under my own observation, where, with the advance of age, the myopia increased ; but, in most instances, the myopia remains about the same through life, unless it be associated with, or be due to,

3

other conditions. This notion, concerning the decrease of myopia with advanced age, may, probably, have arisen from the absurd view formerly entertained by many, that a lack of the fluids of the eye was a cause of the failure of sight, which will be discussed at length hereafter; or, to the view, that flattening of the cornea increased vision for long distances, while it diminished it for short—age favoring a gradual flattening of this organ.

Persons are not always near-sighted in early life, and, in many instances, it is not noticed until they have reached the ages of 20, 25, or even 35 years. And, even then, their first discovery of their myopia is made, when they attempt to use the concave glasses of some friend—distant objects are thereby rendered so distinct that they are astonished, and begin to inquire, if they, too, are near-sighted. They had not previously suspected any defect of vision, because, although they could not see the expression of countenance of an actor on the stage, or a minister in the pulpit, with the same degree of distinctness as others, yet they could distinctly read the finest print, and consequently considered their vision good and not capable of improvement by the use of glasses.

It must be borne in mind, that these differences in the refractive and accommodative powers of the eye are subject to many variations; thus, with some, but one eye only may be short-sighted; or, if the myopia exist in both eyes, it may be much greater in one eye than in the other. Indeed, this is of common occurrence, and one reason why such persons can never procure suitable spectacles, is, because the party, to whom they have applied for glasses, have been ignorant of this most important fact. To remedy this unequal myopia, each eye must be separately tested by concave glasses, or by means of GRAEFE's new and improved optometer.

In many instances, myopia may be partially cured by a methodical use of the eye on distant objects. To a certain extent near-sightedness may be regarded as a habit, originating from a too frequent, or a continued adjustment of the refractive and accommodative power of the eye to near objects; and, as any person may render himself myopic by constant exercise of the eye upon near and minute objects, we may, therefore, with much reason, suppose that a constant exercise of the vision upon objects at a distance, will, effected in a methodical manner and with perseverance, gradually and more or less completely overcome the myopia. And this may be accomplished, not only by looking at objects afar off, but, likewise, when writing, reading, sewing, &c., by adapting the eye gradually, inch by inch, to a longer range.

The question is often propounded to the oculist or the optician, by short-sighted persons, "do you think it best for me to wear spectacles?" In almost every instance, in which myopia is not due to disease, our reply would be, "Yes, if thereby relief is obtained." For in no case should the eye be allowed to over-exert itself, and if vision is assisted by concave glasses, without fatiguing or producing pain in the eye, we should say, by all means wear glasses, except where the myopia is very slight.

In order that the myopic eye may be enabled to see distant objects distinctly, we place before their eyes a suitable concave glass for each eye, which so separates or diverges the rays of light, that they are not brought quickly to a focus in front of the retina, but form the image farther back in the eye, *directly upon* the retina. In figure 3 may be seen the myopic eye; in it we see that the rays do not fall directly upon the retina. We may also see that by placing the concave lens before the eye, the rays are diverged so as to throw the image directly upon the retina, thus giving a clear and well defined image. And the nearer the object is to the eye, the less (or weaker) must be the concavity of the glasses. Hence, near-sighted persons should, invariably, be provided with two pair of spectacles; one for reading at a distance of 10 or 12 inches, (if this can not be done without glasses,) and, for reading music, when playing upon the piano, &c. As a general rule, however, spectacles for reading or writing are not beneficial, and should not be used, except in extreme cases; they are useful, however, for reading music, or, for small objects placed at two or three feet from the eye. The other glass should be a nose-glass of two lenses, adapted for distant objects, and this must not be used constantly, but should be attached to a chain or cord, and worn in some handy place, so that whenever it is required to observe distant objects distinctly, the person may readily be enabled to carry it to the eyes *for the time being only.* This is the true course to be adopted by all near-sighted persons, who desire to benefit their eyes. Many myopes, and especially females, permit their myopia to increase, from a foolish pride, that will not allow them to be seen wearing glasses of any kind; and, hence, they remain in a species of ignorance of the beauties and advantages of distinct distant vision, at the same time, that they are permitting their short-sightedness to become worse.

An error into which near-sighted persons fall, is the

use of a single eye-glass, thus requiring one eye to perform the labor of two, and, as a necessary result, developing one eye much more than the other; this is an injurious practice. Spectacles or nose-glasses are much easier for the eyes and more comfortable for the myope, and should invariably be preferred to the harmful single eye-glass.

In selecting spectacles for myopic eyes, the rule should always be to select the weakest focus for each eye, that will enable the eye to see objects distinctly, without too greatly diminishing or magnifying their size; but, in all cases, it will be much to the advantage of short-sighted persons to consult a competent oculist or optician, and rely on his judgment in selecting the glasses of proper focus. Should the glasses produce a dazzling or glaring appearance, or make objects appear small, they should be exchanged for those of a weaker focus—for glasses should in no way trouble the eye, but should always afford relief. Prof. YOUMANS suggests the following rule for selecting glasses:

"Let the person multiply the distance, in inches, at which he is able to read easily with the naked eye, (say 4 inches,) by the distance at which he wishes to read, (say 12 inches,) and divide the product (48) by the difference between the two, (8), the quotient (6) is the number or focus required."

However, the oculist and optician do not need this rule, as they have more perfect modes of suiting eyes correctly; beside, the rule given, will, frequently, be found to fail in effecting correct results. But when persons are fitted with spectacles by this rule, they should try them for a day or two, and, if found unsuitable, exchange them for others of weaker or stronger focus, as may be required. But by consulting an oculist or optician, all this trouble may be saved, and perfectly correct glasses be furnished.

# CHAPTER V.

## PRESBYOPIA.

PRESBYOPIA, or far-sightedness, is due to a decrease of the range of Accommodation of the Eye, the result of advanced life, whereby the vision of near objects becomes more and more indistinct, while distant objects are seen the same as usual. The outlines of distant objects remain distinct, and no change is perceptible in vision, except in attempts to read or write. It is not a disease any more than gray hairs or wrinkling of the skin, *but is the normal quality of the healthy eye in advanced age.* The popular idea heretofore entertained in relation to far vision, and which still exists to a great degree, is, that as persons advance in life there is a tendency to loss of the fluids of the eye, which organ consequently becomes flattened, and loses its distinctness of vision for near objects. This idea, so frequently found quoted in the most popular text-books of the day, and so generally conceded to be the cause of the failure of vision for near objects, with the advance of age, is incorrect.

Presbyopia, as we observed above, is due to a diminution in the range of "accommodation." * This diminution of the power of accommodation commences in

* The meaning of the term "Accommodation," as applied to the eye, has been explained in Chapter II.

youth at about the age of 10 years, and gradually and steadily progresses as age advances, but is not perceptible until the fortieth or fiftieth year: it is due to a gradual change in the crystalline lens, which increases in firmness during the progress of life. Hence, (supposing, which is not the case, that the ciliary muscles possess the same power as in youth,) the ciliary muscles, even with the same amount of labor as in earlier years, cannot produce the required changes in the form of the crystalline lens for near objects, from which results a diminution of the range of accommodation, and the near point recedes from the eye, so that, although distant objects are seen as distinctly as ever, near ones cannot be seen at the same short distance as in previous years— hence, old persons become what is termed "presbyopic, or far-sighted."

As we advance in years, all the tissues of the system tend to become thicker, firmer, harder and more brittle; the muscles especially become less active, stiff, and more and more inflexible, so that the old man cannot bend his back, nor twist his arm, as readily as the boy. The same law which governs the muscles of one part of the body in this respect, also govern those of all the other parts; and, as the muscles of the eye must undergo this condensing quality effected by age, we may thus understand why the accommodative power of the eye diminishes, the ciliary muscles become less active and powerful, and the crystalline lens becomes much firmer and less yielding to the action of these muscles.

In presbyopia, in consequence of the diminution of accommodation and the proper alteration in the form of the crystalline lens, the rays of light emanating from near objects fall behind the retina, not converging soon enough to form a focus on the retina, and consequently

the images formed upon the retinal layer are indistinct
and confused. But, as the objects are removed from
the eye, the rays emanating from them become less and
less diverging, until having reached a certain distance,
they are brought to a focus directly upon the retina, and
are then seen distinctly. This is the reason why the
presbyope can see distant objects as distinctly as ever,
and with nearly the same degree of sharpness as in
youth, while near objects, as the letters on the page of a
book, or the thread in sewing, become blurred, indistinct, and hardly discernible.

FIG. 4.

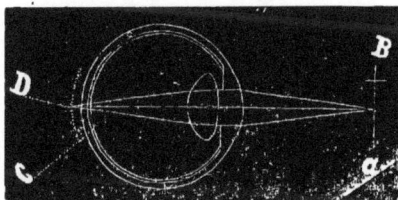

As will be seen in Fig. 4, the converging rays are
brought to a focus *beyond* the retina, (d) and an indistinct and ill-defined image is formed upon the retina, (c)
it will be plainly seen that if the object (a b) be moved
forward the rays will then form directly upon the retina,
(c) but, with the object at a distance of 18 or 20 inches
from the eye, it will be uncomfortable and annoying to
read, write, work, &c. To remedy this, let us put a convex
lens before the eye; now we see the image of the object
(a b) is formed directly upon the retina, (c) and distinct
vision at the proper distance is obtained.

Usually presbyopia occurs between the forty-fifth and
fifty-fifth year; sometimes at a still later period of life,
and frequently in younger years. The latter results

from various causes, as, general debility, premature old age, glaucoma, commencing cataract, &c. Much close work does not essentially injure the eyes in this respect, except where one of the above conditions are present, or, when the work is carried on in an improper light. Myopia is more apt to result from close work. As a general rule, presbyopia may be said to commence when the nearest point of distinct vision lies at about eight inches from the eye—though this does not always involve the use of spectacles. It must be borne in mind, however, that there is another condition of the eye, which we will treat upon hereafter, termed "hypermetropia, or oversightedness," in which, the same as in presbyopia, positive or double convex spectacles are required; and when a person between the ages of twenty and forty has need of convex glasses in order to continue his close work, we will generally find a slight degree of hypermetropia lurking about him.

About the forty-fifth or fiftieth year, or when presbyopia is first perceptible, persons dislike to read small print, as it is less distinct than formerly; vision appears to be inaccurate, especially by artificial light—the letters *u* and *n* are not readily distinguished; the figures 3, 5, and 8 are confounded; a stroke is seen double, a point sometimes multiple. In reading or working by artificial light, the effort becomes painful; near objects become confused, the letters on a page appear blurred, and the reader unconsciously moves the book further from the eyes, until the proper focus is attained, at the same time seeking a bright light. Unless aided by convex glasses, the presbyopic see small or minute objects indistinctly at every distance; because, when near the eye they are out of focus, and when distant they do not reflect sufficient light to make an impression upon the retina. The characteristic mark of presbyopia, and the only true method of correcting it, is by reducing near objects distinctly by means of double convex glasses. Yet how often do we see the far-sighted, in their attempts to read small print or examine fine structures, place the light between the eye and the book or structure, disregarding and trying to escape the necessity of wearing those great assistants, spectacles, and, at the same time, increasing their presbyopia.

So soon, therefore, as accuracy of vision begins to fail, there is need of convex glasses; those of weakest power, No. 60, are at first sufficient, but in proportion as the time of life advances, and the range of accommodation diminishes, stronger glasses will be required. But, from the causes above stated, we sometimes see persons presbyopic at the age of 20, and wearing glasses of the same number as presbyopes 50 years of age; and again, some persons at 50 do not require glasses at all. From these facts it may be readily understood that spectacles are not to be fitted according to the age of the person, but *solely according to the condition of the eye.*

As the firmness of the crystalline lens and the rigidity of the ciliary muscle increases as we advance in years, the range of accommodation also becomes diminished, and the nearest point of vision recedes farther and farther from the eye, and this renders it necessary for the presbyope to change his spectacles every two or three years, each time selecting a stronger power than the former ones used—as the spectacles accurately fitted to the conditions of the eyes to-day, will not answer to the further change effected in these conditions, a year or two hence. "But," says one, "Mr. Jones has worn his glasses for ten years without changing them." True; but examine into the matter and you will find that Mr.

Jones had either hypermetropia, or that at first the glasses were just ten years too strong for him, if I may be allowed this expression, and that by persevering in their use, and resisting the smartings and disagreeable sensations caused by wearing such over-strong glasses, he has succeeded at last *in adapting his eyes to the spectacles*, instead of suiting the spectacles to the condition of his eyes.

Glasses properly fitted to the conditions of the eye, are never uncomfortable, produce no straining or fatigue of the eye, and do not cause headache; but, if these symptoms are produced, the glasses should be at once changed for others that will not occasion them. As the sight is influenced by the intensities of light, a stronger pair of spectacles will be required for artificial light, and if these be of a light blue tint they will with many persons, protect the eye from the deleterious effects of the yellow rays always given out by artificial light. By thus having two pairs of spectacles, one for daylight, and the other for artificial, the eye will be saved a great amount of labor, and its range of accommodation will be much longer preserved.

If, however, a person requires to use stronger and stronger glasses at short and repeated intervals, as two, three, six or nine months, it is an indication of the existence of some serious disease of the eye, as for instance, glaucoma simplex; and, if he wishes to preserve sight, he should at once place himself under the professional treatment of a skillful oculist, as glasses will be of no benefit.

But there are also other conditions of the healthy eye which become more and more apparent as age advances; after the power of accommodation has considerably decreased, a slight diminution of refraction gradually occurs ; 2, a diminution of the acuteness of vision also takes place in advanced age—in consequence of the first, in conjunction with the diminished range of accommodation, the visual lens being parallel, the eye can not be accommodated for distant objects, and the person, usually at the 60th or 70th year, will require one pair of glasses to read or work with, and another pair for walking, seeing objects at a distance, &c. Diminished acuteness of vision, when not the result of disease, should be treated with larger objects instead of larger images. Sometimes presbyopic persons at the age of 85 or 90, have a return of near vision, and are enabled to read, write and sew, without the aid of spectacles.

The proper glasses for presbyopic persons are *convex*, usually double convex ; sometimes *periscopic*, which may be used with advantage, as referred to in another chapter. Convex glasses are numbered according to their focus, thus 60, 56, 52 and 48 inches being the weaker, and 6 and 5 the stronger numbers, and the usual mode of determining the focus of a glass, though not absolutely correct, is to measure the distance at which the glass requires to be placed from a white wall in order to give a distinct image upon this wall, of some distant object, placed, say at 100 or 200 feet off.

The best mode of selecting spectacles is by trying them at the time of purchase. But, persons at a dis‐ tance frequently send to the optician for spectacles, giving him only their age. This is wrong ; the best plan is to inform him at how many inches from the eye they can distinctly read small print. Then the rule for selecting the glasses, as given by the best oculists, is as follows : Subtract the distance at which we desire to place the near point of vision from the actual near point of vision. Thus, if the presbyopic near point be 12 inches, and we desire to find the number of the glasses to bring the near point of vision to 8 inches, we subtract as follows : 1-12—1-8=1-24—convex. 24, is therefore the glass required.  Again, if the presbyopic near point be at 16 inches, and we desire to bring it back to 8 inches, 1-16—1-8=1-15—convex 16, is therefore the glass required.

Presbyopia may exist with hypermetropia or myopia in one eye and presbyopia in the other, and when this is the case, they can refrain from the use of glasses for a long time.   Indeed, as a general rule, when a person of 45 or 50 years can read well without the aid of glasses, he will be found to be thus affected with a combination of presbyopia and myopia.   Or, the same eye may be both presbyopic and myopic; thus, if an eye can see accurately only from 25 to 12 inches, its farthest point of distinct vision is too short, rendering it myopic, and its nearest point at too great a distance, rendering it presbyopic.   If hypermetropia be also present with the presbyopia, glasses must be worn that will correct both these conditions.

The improper use of spectacles in cases of far-sight are of daily occurrence—many interesting cases have come under my own observation, in which the sense of sight has been almost totally destroyed, and again many have believed the eye diseased, in cases where the only trouble was presbyopia.   Presbyopia, as I have said, gradually increases ; many have noticed the gradual failure of vision at 45 and 50 years, but by straining and exerting themselves they have been able to jog along, until at the time when they have applied for spectacles they have been surprised to find that for reading, working, or examining small objects, &c., glasses of considerable power are required.   Such persons very often suppose that they have been suddenly afflicted with "aged" or far-sight, forgetful that every day, gradually but surely, the eye has been undergoing this change, and that every effort of theirs to abstain from glasses has been but another step in the "Deterioration of Vision.'

It is related of MICHAEL ANGELO that in twenty months he completed the gigantic painting of the ceiling of the Sistine Chapel at Rome, and that the effect of the incessant application upon this work rendered him incapable for a long time after, of seeing any picture or near object, except by holding it high over his head—in this case, we have an instance of presbyopia caused by congestion of the eyes.

Presbyopia is often accompanied by neuralgia of the eye, and if the cause of the trouble be not discovered, the patient may suffer severely and be subjected to unnecessary and irksome discipline.

CASE I. *

"A nobleman of exceedingly excitable temperament, consulted me. Neuralgia of the eyes was a source of torment to him. No sooner did he attempt to read a newspaper or a book, but especially the former, the type being more trying, than a painful aching commenced in the eyes and extended to the forehead, causing a profuse discharge of tears, and compelling him to spasmodically close the lids. He discovered that the vexation and worry consequent on this perpetual interruption to a favorite occupation, rendered life miserable. Blisters and a variety of remedies had been tried in vain. Spectacles had never been recommended, as he had a strong prejudice against them. With difficulty, he was per-

* COOPER, on Near Sight, page 92.

suaded to look through my trial glasses, and no prejudice was more effectually or suddenly removed, for he
found that with thirty inch convexes he could enjoy
his newspaper at ease."

I have occasionally met with persons who have used
(abused) their eyes without glasses, after glasses had
become a necessity, and who were thereby rendered
unable to read or write even with the use of glasses.
The eyes had been over-exerted, and even at the time
that they are drawn to seek relief by means of spectacles, these fail to relieve—the eye seems to be partially
blind, and refuses longer to perform its duty. In such
cases, absolute rest for the eyes is required, and the cau-
tious use of the eye for some time on near objects, and
use of spectacles of the proper focus, will eventually
restore vision in a measure, but never so perfectly as it
might have been preserved if even *common sense laws*,
viz: Relief and rest for a wearied member of the body—
relief through spectacles—rest through exercise in the
open air, were obeyed.

### CASE II.

A lady of this city, aged 48 years, called upon me in
reference to her sight which had been gradually failing
for the past four years. Disliking to commence the use
of spectacles, she had delayed their purchase for four
years. I soon found that No. 36 convex was of little
aid; No. 20 convex was little better. With No. 12,
however, she saw comparatively well, but not as good
as she would like to (?); explaining to her that she had
already injured her eyes by their improper use, I
advised rest and quiet for the eye. In two months she
again called upon me, and I found her eyes had very

much improved. With No. 15 she could now read the finest print with ease and comfort.

Similar advice will apply in cases where spectacles or glasses of too *great* a magnifying power have been used, the injury to the eye is the same in both cases. In cases where the eye has been over exerted by the use of magnifying spectacles, four or eight weeks of almost total abstinence from reading and writing either with or without spectacles, *thus* letting the eye have repose, is necessary to alleviate the pain, and in a measure restore the eye to its normal condition, (although the injury can never be entirely remedied.) After four weeks of rest, let the patient commence with the weakest focus with which he can read, No. 2 of the Test Types found in this book. He will soon find himself restored to almost his normal *sight*, and the ill-effects of the improper use of magnifying spectacles partially removed.

CASE III.

Mrs. A——, a lady of wealth, aged 45 years, residing in this city, called on us in November, 1863. A year before, she had found that her sight had already begun to fail. Without any thought upon the subject, she put on her husband's spectacles, No. 20 convex, and wore them for several months. Discomfort in the eyes was soon felt. After reading for a time through these glasses, pain came on, at "first aching in the eyeballs, then a darting, with the sensation as of red-hot sand, in the eyes themselves. The pain soon extended to the brow," with violent head-aches, &c. Rest for two weeks and total abstinence from the glasses, alleviated the pain, and caused the eye to feel more at ease. I then substituted No. 30 convex glasses for the No. 20 she had been

4

wearing, and she was able not only to see the finest print easily, but without any pain or annoyance. Presbyopia is not confined to age; I have seen several cases in which the sight has failed as early as at 12 years.

## CASE IV.

An intelligent girl of 14 was brought to us for aid; her parents, who accompanied her, stated that from infancy she had been accustomed to hold her work off from her; upon examination, it was observed that the distance at which she could read No. 2 Test Types distinctly was 20 inches; by placing No. 22 convex spectacles before the eye, distinct vision was obtained at 12 inches. Although it is not often that we find cases of presbyopia at so early an age, yet at the ages of 18 and 23, we will occasionally find instances of real presbyopia occurring, but care will be needed so that presbyopia shall not be confounded with hypermetropia.

# CHAPTER VI.

## HYPERMETROPIA.

AMONG the several abnormal conditions of the eye, there is none so little understood and so frequently confounded with some other condition, as that which is termed Hypermetropia. Indeed, until within a few years it was wholly unknown, and the cases that were observed were attributed either to a peculiar myopic condition of the eye, or to asthenopia. But, myopia, which has been heretofore described, is an exactly opposite condition, being a short-sightedness due to the focus of the rays of light falling in front of the most external layer of the retina, while hypermetropia is an over-sightedness in which the focus falls behind the most external layer of the retina. On the other hand, asthenopia is a peculiar morbid condition of the eyes, in which, although the eyes are perfectly healthy in appearance, and vision is good, yet they are unable to sustain continued exertion in reading, writing, and upon near work; the objects becoming, after a time, confused and indistinct, with a sense of fatigue in the eyes.

Again, hypermetropia has been often confounded with presbyopia, and the term formerly applied to it, "hyper-presbyopia," indicates that it was considered a condition or degree of presbyopia; but the presbyopia of advanced years is due to a diminution of the range of accommodation, the near point of vision only being removed too far from the eye, while distant objects are seen as well as ever; while in hypermetropia, vision is never accurate either for distant or for near objects, and every effort to distinguish any thing, to which is united great knitting of the brow, is rapidly followed by fatigue, especially among those whose hypermetropia is considerable.

J. SŒLBERG WELLS, a celebrated London Opthalmologist, in writing of this condition, observes : " By hypermetropia is meant that peculiar condition of the eye, in which the refractive power of the eye is too low, or the optic axis (the antero posterior axis) is too short; we may, however, have both these causes co-existing.   To illustrate, let us refer to

FIG. 5,

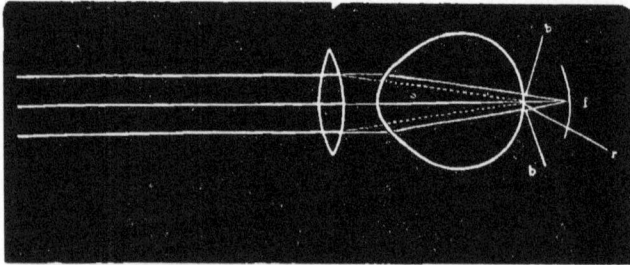

which represents a hypermetropic eye, " in which, as we have before said, either on account of its being too short in the antero posterior axis, or of its possessing too low a power of refraction, parallel rays are brought to a focus, not upon the retina (r), but behind it at (f); circles of diffusion (b b) are formed upon the retina, and the object consequently appears blurred and indistinct.

From this it is easily seen that the less the power of refraction in the hypermetropic eye, the greater the effort of the accommodation. "By placing a convex lens before the hypermetropic eye, the parallel rays are made so convergent as to fall directly upon the retina (r), without any effort of accommodation, and we thus place the hypermetropic in the same condition as the normal eye, upon whose retina parallel rays are united without almost any effort of accommodation."

We may produce an artificial hypermetropia, by placing concave glasses before the eyes, which will be followed by an immediate indistinctness of vision, and a sense of discomfort to the eyes, because the rays emanating from objects are not brought to a focus upon the retina, but behind it, and the image formed upon the retina must necessarily be confused and misty. Now, keeping the concave glasses before the eyes, if we place before these the proper convex glasses, distinctness of vision will be restored. This is exactly what occurs in hypermetropia.

A hypermetropical eye is generally small and imperfectly developed; immediately around the cornea, the sclerotica has a flat, slightly curved appearance, while in the equatorial region of the eye the curvature is strong, being, however, much greater in the direction of the meridian than in that of the equator itself. The eyelids are flat and broad, and the eyes are far from one another; sometimes the eyes are deeply seated, at others, superficial, or apparently full and large; however, nothing can be inferred from these positions of the eyes. I have frequently adapted glasses for hypermetropia, especially among children, in whom the eyes presented a full and prominent appearance. In the more severe degrees of hypermetropia, there will generally exist, in addition to the above, a want of full development of the bones of the face.

It has already been stated that the indistinctness of vision in hypermetropia is caused by the refractive power of the eyes being so low that parallel rays, when the eye is in a state of rest, fall behind the retina, and only convergent rays are brought to a focus upon it. This is exactly the opposite of myopia, in which we place concave glasses before the eyes, in order that a divergent direction may be given to the rays. While, in hypermetropia, we substitute convex glasses for the concave, and thus give a convergent direction to the parallel rays, instead of a divergent.

If the hypermetropia is absolute, the hypermetrope will experience great difficulty in seeing distant objects, and, for close work, he will be almost blind. In reading, writing, or at close work, he will invariably carry . the book or work near to the eye—and owing to this action, the difficulty is often attributed to short-sightedness, from an ignorance of the real defect. In those cases, however, where the hypermetropia is slight, no particular inconvenience will be experienced in looking at distant objects; but when an attempt is made to read or write, the eye wearies from the continual straining required, and continually the book or paper is put aside.

It is frequently the case, that a hypermetrope jogs along unconscious of any defect in distant vision, and attributing such defect as he has observed, either to a weakness of the eyes, or to a myopia that can not be remedied by glasses. If he applies to an oculist or an optician, who is thoroughly versed in these conditions of the eyes, and who places before his eyes the proper convex glasses, he is surprised to see what new fields of observation are opened before him, and then for the first

time learns how blind he had previously been; and he
rejoices to find himself thus enabled to distinctly see and
examine the leaves of the tree, as well as other distant
objects, and he occupies himself in gazing at these mas-
terworks, now spread out before him like a picture in
bold relief, all of which were previously mere dead letters
to him. Nor is this all : he finds that the convex glasses
also enable him to read, write, and perform close work,
without the eyes becoming fatigued.

But, hypermetropia may co-exist with presbyopia.
" If, for instance, we find that a person suffering from
hypermetropia can not read with the glasses which
neutralize his hypermetropia nearer than twelve or fourteen inches, he suffers at the same time from presbyopia,
and will require two sets of convex glasses : One pair for
seeing objects from fourteen inches to a distance, and a
stronger pair for objects nearer than fourteen inches."

It is unfortunate for those laboring under defects of
vision, that many of them allow pride to master common
sense, and will not, therefore, make use of the only
means—spectacles—that will restore distinct vision to
them.

### CASE I.

A lad, ten years of age, applied to me one day, in
order to ascertain whether spectacles would benefit him.
His parents thought him nearly blind, and wished to
know if his vision could be improved by glasses. With
No. 6 convex, the boy was enabled to see clearly and
distinctly; and yet, though restored from comparative
blindness to perfect sight, and notwithstanding they had
all desired to ascertain what service could be had from
glasses, it required considerable argument to overcome

pride and induce the lad and his parents to consent to
the constant use of spectacles. And from this cause,
perhaps,.also, from others, as the foolish persuasion of
friends or parties who are wholly ignorant of these matters, hypermetropes will often neglect the true and
only remedy, and wander on in darkness, even while
light is being constantly offered to them. Many of the
cases of squinting or strabismus that we meet with, are
the results of neglected hypermetropia, for it is now a
well known fact, that hypermetropia will occasion strabismus.

I can not too strongly advise every person laboring
under hypermetropia to consult a competent oculist or
optician, in order to have glasses properly fitted to the
condition of his or her eyes, and thus save their sight.
I would also call the attention of parents and teachers to
this matter. There have been many instances of children being punished at school because of delinquency in
committing lessons, while at the same time, argues the
irritated teacher, "you can play out in the yard, and
run about well enough with the other scholars." And
on this account he concludes that if the child is able to
play, it is also able to study. Preceptors and parents
can not be too careful in this matter; the child may be
hypermetropic, and if this be the case, it is an impossi-
bility for him to study, at least until a suitable pair of
spectacles have been adapted to his eyes. With them
he can study and commit his lessons to memory; without them he can do nothing. With them his sight will
be preserved; without them, vision will become more or
less impaired, or a convergent strabismus may take
place.

Many people entertain a prejudice against wearing
spectacles, supposing they are injurious to the eyes.

This is a great mistake, and has undoubtedly arisen from observing the effects of improperly selected glasses upon the eyes of those wearing them. Persons who are myopic, presbyopic, or hypermetropic, and who wear properly selected glasses, always have their eyes improved thereby, and never, in any instance, can the slightest degree of injury follow. But, be sure, when you apply for spectacles, to consult only with well known oculists and opticians.

DONDERS relates the following case in his "Accommodation and Refraction of the Eye," page 285:

### CASE II.

The Rev. G. D., aged 52, looks dejected. "My good Professor, I come to you, for I feel that I am getting blind." For the last twenty years he has thought that within a year he should be blind, and singularly enough, although he still sees, he continues to look upon every year as his last. Such is the man. His life has been a struggle with his eyes. Even as a child he read with difficulty. When a student the least exertion fatigued him, and he was compelled to learn more by hearing than by his own study. As a preacher, he has been obliged to write his sermons in a rather large hand, and still get them by heart. And what was the worst part of it, he never read nor worked without the idea that he was thus hastening his blindness—interfering with the concentration of his mind upon any definite object. The same fear of blindness had restrained him from a matrimonial alliance with which he believed his happiness for life to be connected. He trusted in art. He had faith in a person. He consulted in Germany, and if the optician had sometimes given him spectacles which had brought him relief, these were mercilessly taken from him again by oculists, on the first consultation, as a treacherous instrument, which must in the end inflict upon him the total loss of his sight. At last in his fortieth year he got convex glasses of No. 40, and he now uses No. 20. "Do you see with these spectacles at a distance?" was my first inquiry. "Something better, but still imperfectly." I tried No. 10. "Much better," was his verdict. Subsequently I gave No. 8; "still better." He got No. 7 to wear. The man was as grateful as a child. He left me as one saved from destruction.

Such victims of prejudice against the use of convex glasses, are not uncommon. On the contrary, many to whom the light is once offered, or to whom the avenue to vision is offered, grasp at it as a drowning man at a straw.

CASE III.

A well known physician, of this city, while yet a boy of 16 or 17, was unable to see, except in a dazzling light—even then the smaller letters were indiscernable and blurred. By holding the book close to the face, he was enabled to see the larger letters. Having been in this state since birth, his parents and friends supposing him blind, left him to his fate—blind for life. What a sad reflection, to be cut off from all intercourse with the material world—he did not ever expect to be able to see any better. At one time on an errand to the village store, he came in as the store-keeper was fitting a lady with a pair of spectacles. Picking up a pair, he put them on. They were No. 10 convex. With them, he found he could see very much better, that small letters were made plain, that a new "World" was opened to him.

He immediately purchased the glasses. Since then, he has been fitted by opticians, gradually, degree by degree, increasing the focus, until now, with No. $4\frac{1}{2}$, which he wears, vision is restored to him as perfectly as could be desired.

# CHAPTER VII.

### CATARACT.

CATARACT is an opacity of the crystalline lens, or of its membranous envelop or capsule, obstructing the rays of light, and rendering vision so impaired as to amount to little or nothing more than a perception of light from darkness. A person afflicted with cataract (the retina being healthy), never becomes totally blind; he is always able to distinguish day from night, but is unable to distinguish objects around him.

Cataract is generally a slow, harrassing disease, gradually and surely growing upon the patient At first, there is a diffused mist or fog before the eye, which gradually thickens, and becomes denser. At the early stage of cataract, the film or mist seems so slight that no solicitude is felt—but as the cloud thickens to such an extent that vision is obstructed almost entirely, the patient becomes painfully sensible of the loss he has sustained. It must be recollected that the dark spots frequently seen floating before the eyes (*muscæ volitantes*), are not a symptom of cataract, and are more commonly of but little importance; while those that are fixed, always remaining in one spot, are indicative of some unhealthy condition of the retina (*amaurosis*), that may eventually terminate in complete, and even incurable blindness.

Is cataract always a slow progressive disease? RICH-
TER relates the case of a patient laboring under gout,
exposing his feet to a great degree of cold during the
night—the gout retroceded, and he was deprived of
sight. RICHTER saw him next morning, and found a
complete pearl-colored cataract.

In the Dublin Medical Press, of May 4, 1842, it is
stated that Dr. MARTIN, of Portlaw, at a meeting of the
Surgical Society of Ireland, related two cases of sudden
formation of Lenticular Cataract. The one occurred in
a cachectic woman, who, after setting up for several
nights with her invalid mother, and crying a great deal,
awoke one morning with her crystalline lenses semi-
opaque, and presenting the appearance of being stellated
from the centre, as if breaking up during maceration.

The other was in a man, who, having been married to
a farmer's daughter, retired to bed after the usual fun
of an Irish wedding, his eyes being perfect, and awoke
early in the morning, with his sight greatly impaired
from cataracts. MACKENZIE says "He has not wit-
nessed any such sudden formation of cataract in eyes,
previously sound." And, indeed, when they do occur,
they may be considered as exceptions to the general
rule, being due, probably, to the existence of some previous, but unnoticed, affection of the crystalline or its
capsule, or to some peculiar constitutional depravity.

When cataract exists from birth, it is termed *congenital cataract*, and may be derived from progenitors who
were subject to it, or be due to changes occurring in the
lens either previous to, or immediately after, birth, independent of any hereditary tendency. And this form
of cataract may frequently exist until the age of puberty,
without giving rise to any annoying symptoms, in which

respect, as well as in its conditions and growth, it differs essentially from the cataract of elderly persons.

WILLIAMS (of Boston), in his "Diseases of the Eye," page 143, describes the appearance of the lens in congenital cataract as showing a "grayish or bluish white opacity, sometimes perfectly uniform, like milk diluted with water, in other cases mottled with specks of a chalky white." In many cases, the opacity is confined at first to the centre of the lens, but spreading, it soon obscures vision entirely.

In cases of blindness from birth, the twitching or oscillation of the globe of the eye is very noticeable, and is due to the patient never having learned to control the direction of the eye.

WILLIAMS relates that he operated in one day upon the six eyes of three children in a family where yet another child and the mother were also affected—two other children remaining thus far exempt. The result was perfectly successful in all the eyes operated on.

The person upon whom a cataract is formed, sees by far the best by twilight or in a partially darkened room, while by a bright light his vision is more or less indistinct. The reason of this is, that in a moderately dull light, the pupil becomes dilated so that more of the crystalline lens is exposed, and the rays of light pass through the less opaque border of the lens, and fall upon the retina; while in a bright light, the pupil contracts, and thus covers the less opaque portion of the lens, thereby depriving the retina of the action of the rays of light, which are more or less intercepted by the greater central opacity of the lens. From this known fact, charlatans, often professing to perform great cures for "Cataract," only apply belladonna to the eye, which expands

/

the iris, and by thus enlarging the pupil, causes a momentary improvement of vision.

The following table exhibits the relative frequency of cataract at different periods of life. FABINI treated 500 cataract patients :

Males,..................................................................268
Females,..............................................................232

| | | | |
|---|---|---|---|
| From | 1 to 10 years, | .......................... | 14 |
| " | 11 to 20 " | .......................... | 16 |
| " | 21 to 30 " | .......................... | 18 |
| " | 31 to 40 " | .......................... | 18 |
| " | 41 to 50 " | .......................... | 51 |
| " | 51 to 60 " | .......................... | 102 |
| " | 61 to 70 " | .......................... | 172 |
| " | 71 upwards, | .......................... | 109 |

500

From the above it will be seen that the majority of cataract cases occur after the fortieth year. The disposition to this disease being small before that time, but increasing as life advances.

Cataract is often produced by injuries; a rupture of the capsule of the crystalline has been known to develop opacity of a considerable part of the lens, "in four-and-twenty hours."

MACKENZIE gives us ten remote and predisposing causes to cataract, classing them as follows :

" 1. Old age.

" 2. Hereditary tendency.

" 3. Exposure to strong fires, as blacksmiths, glass-blowers, cooks, laundresses, &c.

" 4. Sedentary habits and intense looking at objects while in motion, as the stocking weavers, cotton workers, &c.

"5. The use of wine and spirituous liquors, but especially the former, appears to favor the production of cataract, which is a common disease in all countries where wine is so cheap, as to be the habitual beverage of the lower order.

"6. Constant use of opium, hence the frequency of this disease among the Turks.

"7. The inhabitants of volcanic countries, as Naples and Sicily are said to be very subject to cataract.

"8. The sudden application of cold to the extremities of the body, so as to check any natural or morbid effort or evacuation, such as menstruation, or a paroxysm of gout, is apt to produce cataract.

"9. As a general rule, the subjects of cataract enjoy good general health. They complain more frequently of rheumatic affections, than of any other; dyspepsia, pains in the head, and giddiness not unfrequently precede cataract, especially in women.

"10. Cataract is frequently the result of diabetes mellitus."

An ounce of prevention is greatly better than a pound of cure—and, knowing the *causes*, or, we may say, the *predisposing* causes to cataract, we may, by avoiding these, as much as possible, prevent or retard the formation of a cataract. But after the cataract has become fully formed, and so far advanced as to positively render one blind, sight can only be restored by an operation, which has, for its object, the removal of the opaque lens from in front of the retina, so as to allow the rays of light to fall upon this membrane. The character of the operation depends somewhat on the kind and character of cataract; but generally the lens is removed either by depression, or by extraction. By the first operation, the lens is pushed aside and downward, so as to be entirely out of the way of the pupil; this is not always a successful operation. The latter operation is the entire removal of the lens from the eye, through a large flap made in the cornea, or, by a more recent one of GRAEFE's, in which the cut is made through the sclerotic and outside of the margin of the cornea. Other operations are also performed for cataract, but as surgical operations are not included within the scope of this work, we cannot dwell upon them. We will only say, that great care should always be taken to place the patient to be operated upon, in the hands of a skillful, responsible, and well known surgeon or oculist; such an one may be successful in restoring vision, while a novice or an impostor will be very apt to ruin it forever. For it must be borne in mind, that all operations for the removal of cataract are not successful, a greater or less risk of total loss of the eye and permanent blindness, attends them all.

" During the year 1830, the autumn of 1832, and the spring of 1833, Professor ROUX operated by extraction, on 115 patients, and 179 eyes, at the *Charite*, in Paris, with the following results," which furnishes us with a correct idea of the proportion of successful operations in cases of cataract—the proportion being somewhat more than 5 out of every 8 patients upon whom he operated successfully:

73 patients recovered sight, viz: - - $\left\{ \begin{array}{l} \text{40 men} \\ \text{33 women} \end{array} \right.$

97 operations succeeded in - - - $\left\{ \begin{array}{l} \text{52 men} \\ \text{45 women} \end{array} \right.$

72 operations failed in - - - $\left\{ \begin{array}{l} \text{32 men} \\ \text{40 women} \end{array} \right.$

10 partially succeeded in - - - $\left\{ \begin{array}{l} \text{6 men} \\ \text{4 women} \end{array} \right.$

The operation of GRAEFE, referred to above, has, it is

5

stated, proved more commonly successful than any other
that has yet been undertaken for the removal of cataract, besides being attended with much less risk of permanent blindness.

After the removal of the crystalline lens, objects are
formed but imperfectly on the retina, because of the
refractive power of the eye being considerably diminished.

We must, therefore, in order to properly converge the
rays of light to focus upon the retina, and thus render
vision *perfectly distinct*, substitute a biconvex, plano-convex, or meniscus lens on the outside of the eye, in place
of the removed biconvex crystalline. But, as the accommodative power of the eye is now almost wholly lost, it
would require an almost infinite number of lenses to
give *perfect vision* at any distance. This, however, is
impracticable, hence this class of persons are usually
furnished with two pairs of spectacles, one for reading,
writing, &c., and the other for walking and vision at a
moderate distance.

"Cataract spectacles," as they are termed, vary in
power from 4½ inch to 1 inch focus (see chapter XI).
In cataract, as in other difficulties of the eye requiring
spectacles, the weakest powers are the ones that should
be chosen. Oval frames of steel are to be generally preferred; but where the patient is a farmer or mechanic,
strong silver or German silver frames are preferable, as
they will stand a great deal of hard usage. Spectacles
should not be used by cataract patients as long as they
can see distinctly, or as long as their vision continues to
improve, without their use; neither should they be used
too soon after an operation. With regard to these points,
too much care cannot be taken, and I would especially
call attention to the following remarks by Dr. MACKEN-

ZIE: "The too hasty employment of cataract glasses after the most successful operation, may soon bring the eye to a state of weakness, which will render it unfit even for those employments which require but a moderate degree of sight. No cataract glasses ought to be given to a patient so long as his vision appears to be improving without their use. This generally continues to be the case for several months after the operation. If we allow our patient to use cataract glasses during this period, he will, no doubt, be very glad to find that he can return immediately to almost all his ordinary pursuits; but he will soon begin to observe that he does not see so well as he did, and this he will probably remedy by a new pair of glasses of greater convexity, and he will go on in this way changing his glasses as his power of vision becomes less, till at last he ends in finding none which enable him to see so well as he did with those which he first used. On the other hand, if our patient does not begin to try cataract glasses till he has completely recovered from the operation, and the eye has as much as possible habituated itself to the absence of the crystalline lens, if he then select proper glasses and use them for a while only occasionally, his sight will still continue to improve, and his first glasses will, probably, if he be an old man, serve him for his life; and if he be a man of 30 or 40 years, he will not require to change them till he be 50 or 60. The operative means for the cure of cataract may have perfectly succeeded, but from want of proper glasses, the patient may derive but a small amount of benefit."—The same rules apply in fitting cataract patients, as in fitting spectacles for the natural defects of vision, and which will be further discussed in another chapter.

# CHAPTER VIII.

## ASTIGMATISM.

THAT peculiar condition of the eye to which the term *astigmatism* is applied, is by no means uncommon. The normal eye itself possesses a certain degree of astigmatism; but it is only when it exists in excess that it becomes a disease. By the word "astigmatism," we understand a difference of refraction of the rays of light that pass through the various meridians of the eye, and which occasions more or less disturbance of vision, so that a point appears elongated into a line, a circle appears oval, a square presents the form of a parallellogram, letters appear distinct in one meridian, and blurred or indistinct in another, &c. &c. Out of every hundred cases of patients affected with disturbances in the refraction of the eye, about three are astigmatic; it is more frequent among males than females.

By *meridians* of the eye, we understand the planes or imaginary straight lines that intersect the optical axis in various directions; this axis being represented by a line passing through the center of the cornea to the most distinct point of vision upon the retina of a healthy eye. The plane that intersects this axis vertically is called the "vertical meridian," and that which crosses it horizontally is called the "horizontal meridian." In order to have clear and distinct vision, all the rays of light from any point of an object looked at must unite in a point upon the retina; but, it must be recollected, that some of these rays pass through the vertical meridian, others through the horizontal, and so on. If the refraction in the vertical meridian is much greater than that in the horizontal, the rays passing through the former will become united much more speedily than those passing through the latter, and the image of the point of the object formed upon the retina will be indistinct, because the rays in the vertical meridian unite in a single point upon the retina, while those in the horizontal meridian unite either in front of, or behind, the retina, forming upon this *circles of diffusion*. Prior to the researches of DONDERS, but little was known concerning astigmatism; true, THOMAS YOUNG first observed this condition of the eye (normal astigmatism), in 1793; and, in 1827, AIRY, Royal Astronomer, first investigated this subject as far as related to his own eye, which was astigmatic. STOKES was the first who undertook to determine the degree of astigmatism by cylindrical lenses. GOODE, HAMILTON, HAYS, and a few others, had reported some cases; but it was not until the subject had been carefully investigated by DONDERS, that correct views upon the difficulty became generally known. Astigmatism is more generally congenital, though it may be acquired; sometimes both eyes are equally affected; at other times, one eye alone will be affected, the other being perfectly normal. When it is *congenital*, it may be due to an abnormal situation of the crystalline lens, but more frequently to a particular conformation of the cornea, which does not present the same radius of curvature in its various meridians. From the measurements made by DONDERS, KNAPP, and other eminent oculists, the radius of curva-

ture in the horizontal meridian of the cornea is found to
be, more frequently, much longer than that in the vertical meridian—or, in other words, the *focal distance* of
the cornea is, generally, shorter in its vertical than in
its horizontal meridian. Astigmatism may also be due
to an obliquity of the cornea with the optical axis; so
that instead of the corneal axis intersecting the optical
axis at right angles, it forms with this axis an angle of
nearly 6 degrees.

*Acquired* astigmatism may be the result of inflammation of the cornea, in which the shape of this membrane
becomes changed; or, it may be due to an oblique position of the crystalline lens, effected by a blow or other
injury.

Persons laboring under astigmatism, more commonly
have hypermetropia also; the features are apt to be flat;
the anterior surface of the sclerotica presents a feeble
curvature, and a sudden incurvation in the vicinity of its
equator. The optical axis is short; the anterior chamber shallow; the pupil quite small; a slight squint outwardly is often observed; and, with many, the vertical
axis of the eyeball is perceptibly smaller than the horizontal; the cornea is either shorter than usual in its vertical meridian, or it extends farther backwards, from
increased curvature, so that the section between the cornea and the sclerotica does not lie in the same plane.

The acuteness of vision is diminished, or has even
been thus from infancy; vision itself is more or less confused, and, as already stated, points appear as lines, circles as ovals, squares as rectangles, and when lines
intersecting each other at various angles are looked at,
some will be seen much more distinctly than others, and
if they are all of equal length, some will appear longer
than others, &c. &c.

Astigmatism may be detected in various ways. If a circle, or a circular orifice, is presented to the party to look at or through, it appears oval to him; if it be a square, it appears rectangular. If a number of straight lines be placed in two rows close to each other, one row having the lines running vertically, the other horizontally, the lines being at equal distances from each other, and held at a certain distance from. the eye, the patient will see, according to the direction of his astigmatism, one row quite distinctly, and the other very confusedly. Again, if a slit about one-third of a line in width, and six lines long, be made in a box cover, or in a disc of brass, silver, or other material, and then be held before the eye to look through, it will be found that by slowly revolving this disc before the eye, so that the slit will alternately be in parallellism with the various diameters of the cornea, there will be more distinctness of vision when this stenopaic slit corresponds with certain diameters, than with any of the other diameters of the cornea. As astigmatism more generally exists associated with myopia or hypermetropia, concave or convex glasses will often be required in these examinations, in order to improve the patient's sight.

It is not my design to enter into a professional description of causes, symptoms, diagnosis, and treatment, which would enlarge the work beyond the limit intended for it; the principal object in view is to post the public as to the common causes of the difficulties of vision so frequently met with, and to explain how and where they can have such difficulties remedied, and thus protect themselves from being imposed upon by mere pretenders and ignorant empirics. Hence, I shall not enter into an explanation of the instruments used in the diagnosis of astigmatism.

Our best oculists are provided with the necessary opthalmoscope, stenopaic apparatus, GRAEFE's new optometer, SNELLEN's test type, concave, convex, and cylindrical lenses, &c. &c., for the diagnosis of astigmatism, as well as for the determination of the particular kind of cylindrical glasses suitable for any given astigmatic eye; and, I therefore advise all parties whose vision can not be corrected by the usual concave or convex glasses furnished by the optician, to consult with any oculist of good standing and experience, who will readily determine the character of the glasses to be worn, in order to render vision clear and distinct. And, as the cylindrical glasses are more generally imported from PAETZ & FLOHR, Berlin, it may require eight or ten weeks after the examination before they can be had.

In order to convey a more thorough idea of this difficulty of vision, I will give a few quotations: Prof. AIRY, referred to above, "discovered that, in reading, he did not usually employ his left eye, and that in looking at any near object, it was totally useless; in fact, the image formed in that eye was not perceived, unless attention was particularly directed to it. Supposing this to be entirely owing to habit, and that it might be corrected by using the left eye as much as possible, he endeavored to read with the right eye closed or shaded, but found that he could not distinguish a letter, at least, in small print, at whatever distance from his eye the characters were placed. Some time afterwards, he observed that the image formed by a bright point, such as a distant lamp or star, in his left eye, was not circular, as it is in the eye which has no other defect than that of being near-sighted, but elliptical, the major axis making an angle of about 35° with the vertical, and its higher extremity being inclined to the right. Upon putting on concave spectacles, by the assistance of which he saw distant objects distinctly with his right eye, he found that to his left eye a distant lucid point had the appearance of a well defined line corresponding exactly in direction, and nearly in length, to the major axis of the ellipse above mentioned. He found, also, that if he drew upon paper two black lines crossing each other at right angles, and placed the paper in a proper position, and at a certain distance from the eye, one line was seen perfectly distinct, while the other was barely visible: while upon bringing the paper nearer to the eye, the line which was distinct, disappeared, and the other was seen well defined."[*]

Again. " Dr. ROBERT HAMILTON relates the case of a patient who, when looking at a clock, was unable to distinguish the hands if they pointed perpendicularly, as at six o'clock, but if horizontally, he had no difficulty; so when looking at a wheel at a little distance, the horizontal spokes only could be seen. The patient was a coach painter by trade, and this peculiarity of vision greatly interfered with his business, for he could not draw vertical lines with any degree of correctness, and unwittingly made them slanting, a serious fault in heraldic devices; horizontal lines he drew with perfect precision."

DONDERS classifies astigmatism into three forms, as represented in the annexed table, 1, 2, 3, being subdivided each into classes, a, b :

---

[*] Mackenzie, p. 867-8.

| (1) | (a) | (b) |
|---|---|---|
| Simple astigmatism. | Simple myopic astigmatism. | Simple hypermetropic astigmatism. |

| (2) | (a) | (b) |
|---|---|---|
| Compound astigmatism. | Compound myopic astigmatism. | Compound hypermetropic astigmatism. |

| (3) | (a) | (b) |
|---|---|---|
| Mixed astigmatism. | Mixed astigmatism with predominant myopia. | Mixed astigmatism with predominant hypermetropia. |

I. The state of refraction of one principal meridian* is normal—the other either myopic or hypermetropic.

(a). One principal meridian myopic—the other normal.

(b). One principal meridian hypermetropic—the other normal.

II. The myopia or hypermetropia exist in both principal meridians, but it varies in degree.

(a). Both principal meridians—myopic.

(b). Both principal meridians—hypermetropic.

III. Rarely found—one principal meridian myopic—the othe hypermetropic.

(a). Myopia the predominant feature.

(b). Hypermetropia predominant feature.

---

*The vertical and horizontal meridians, are termed the "principal meridians;" though, astigmatism may exist in any of the meridians.

Astigmatism is often congenital. It occurs oftener in hypermetropic eyes than in any other. DONDERS places the proportion of hypermetropes also astigmatic as 6—1.

Astigmatism may be remedied by cylindrical lenses which correct the irregularities of refraction. As the astigmatism is the result of an inequality of curvature in the vertical and horizontal diameters of the cornea, in order to produce distinctness of vision, the refraction of the various meridians must be corrected, by artificially increasing or diminishing their curvature. A cylindrical lens produces this effect only for a single diameter, that is to say, it causes the rays of light to converge or diverge, following the diameter perpendicular to the axis of the lens, while it has no effect upon the rays that pass parallel to the axis; or in other words, it refracts only those rays of light the strongest which strike it in a plane at right angles to the axis of cylindrical curvature. In order to correct compound astigmatism, positive and negative *sphero-cylindrical* and *bi-cylindrical* lenses are employed.

Professor AIRY, to remedy his defect, "made a pin hole in a blackened card, which he caused to glide on a graduated scale; then strongly illuminating a sheet of paper, and holding the card between it and the eye, he had a lucid point on which he could make observations with ease and exactness. Then resting the end of the scale on the check bone, he found that the point at the distance of 6 inches appeared a very well defined line, inclined to the vertical about 35°, and subtending an angle of 2°. Again, at the distance of 3½ inches, it appeared a well defined line at right angles with the former, and of the same apparent length. It was, therefore, necessary to make a lens which, when the parallel

76 THE HUMAN EYE.

rays were incident, should cause them to diverge in one plane from the distance of $3\frac{1}{2}$, and in the other plane from the distance of 6 inches. The professor obtained a lens of which the radius of the spherical surface was $3\frac{1}{3}$ inches, of the cylindrical $4\frac{1}{2}$ inches, and with this he was able to read the smallest print.

In Dr. HAMILTON's patient, the relation of the horizontal to the vertical focus appeared to be as $5\frac{1}{2}$ inches to $6\frac{1}{2}$ inches, and on trying him with plano-concave cylindrical lenses, it was found that a lens of 24 inches focal length, the cylindrical surface being made to act horizontally, operated very beneficially."

For particulars as to fitting, and various kinds of cylindrical glasses, refer to chapters XI and XII.

# CHAPTER IX.

## STRABISMUS.

It was not my design, originally, to have entered upon a description of that deformity of the eye known as strabismus or *squinting*, as it does not strictly belong to a work of the present character, and I have only consented to devote a few pages to its consideration at the request of many of my friends. Strabismus is a condition of the eye in which there is a want of harmony between the two optical axes, a deviation of the optical axis of one eye not only in its relation to the optical axis of the other eye, but also in regard to its own orbit, so that, in attempting binocular vision there is a defect in the convergence of the optical axes towards objects looked at; one eye, involuntarily and without regard to the movements of the other, turns away from its natural direction. In normal binocular vision the optical axes are directed towards, and converge upon, the object looked at, so as to form an angle the apex of which corresponds to the object; this convergence of the optical axes causes the image of the object to be formed upon corresponding points of the two retinæ, and thus assures *simple* vision with two eyes. The true optical axis is a line passing from the yellow spot of the retina, (*macula lutea*) through the center of the crystalline lens and the

center of the cornea. Any defect in this exact convergence occasions double vision (diplopia) and strabismus.
Very frequently a squint-eyed person sees double previous to the occurrence of the squint; but as the natural
tendency is to simple vision, the offending eye is gradually deviated by certain of its muscles, until vision is
performed with but one eye.

There are some instances in which strabismus is due
to a paralysis, an insufficiency of action of certain muscles; or, perhaps, to an excess of action of one of the
muscles of the eye, generally the internal rectus on one
side, and this excess of action may be either relative or
absolute. Sometimes, the strabismus is hereditary; at
other times it is mechanical, due to tumors as well as to
what is termed "posterior staphyloma," or atrophy of
the choroid, at its insertion around the optic nerve.
DONDERS has thrown much light upon this subject; he
has ascertained that by far the greater number of cases
of *convergent* strabismus are due to an existing *hyperme-*
*tropia;* while those of *divergent* strabismus are almost
always the result of *myopia.* Yet it does not follow that
every hypermetropic or myopic person should squint,
for we meet with many of these classes without any strabismus whatever; and, again, convergent strabismus is
by no means uncommon among myopes. DONDERS
observes as follows:

"Experience, in the first place, shows that strabismus
convergens is in the great majority of cases combined
with hypermetropia. In 172 cases investigated by us,
hypermetropia was 133 times proved to exist in the
undeviated eye." "The hypermetropic individual must,
in order to see distinctly, accommodate comparatively
strongly. This holds good for all distances. Even in
looking at remote objects, he must endeavor to overcome

his hypermetropia by tension of accommodation, and in proportion as the object draws near, he must still add so much accommodation as the normal emmetropic eye should need on the whole. The vision of near objects especially requires extraordinary tension. There exists a certain connexion between accommodation and convergence of the visual lines; the more strongly we converge the more powerfully can we bring our faculty of accommodation into action. A certain tendency to increased convergence so soon as a person wishes to put his power of accommodation upon the stretch, is therefore unavoidable. This tendency exists in every hypermetropic person." "Hypermetropia is a very wide-spread anomaly. I am convinced that it occurs more frequently than myopia."

In cases of very high degree of hypermetropia, strabismus is rarer than in weaker or lower degrees. There are other causes promoting squint, such as constant looking at a side object; a spot upon the nose; various diseases of childhood, etc., etc. There is also occasionally met with, an apparent strabismus in cases of simple vision, due to the yellow spot of each retina not being situated at corresponding points; thus, the optical axis of one eye passes to the yellow spot, while, in the other eye, the spot is inside or outside of this extremity of the axis. GRAEFE terms this *incongruence of the retinæ*.

Strabismus is designated according to the direction in which the faulty eye deviates. It is termed *convergent, internal*, or *nasal*, when the eyeball turns inwardly, looking toward the nose; *divergent, external*, or *temporal*, when the eyeball turns outwardly, from the nose; *sursumvergent, superior, upward*, or *frontal*, when the eyeball turns upwardly; and *deorsumvergent, inferior, downward*, or *jugal*, when the eyeball turns downwardly. When the globe is not turned directly in any of the directions just referred to, but in an intermediary direction, the strabismus is then termed *oblique;* it may be upward and outward, or, downward and inward. When one eye is turned in one direction, and its fellow in another, it gives such an expression to the face that it has been termed *horrible* strabismus. The most common variety of strabismus is the *convergent.*

When the healthy eye of a squinting person is covered with the hand, or with a blind, and the affected eye resumes its proper normal position, the strabismus is then termed *active, continuous, spasmodic,* or *permanent;* but if the squinting eye still preserves its original position, notwithstanding the efforts of the patient to bring it into a proper direction, the strabismus is then termed *paralytic, fixed,* or *passive* strabismus; it is also termed *luscitas,* and is usually due to paralysis, tumors, or to false growths or bridles, the consequence of a previous suppurative inflammation. The term *anchylosis of the eye,* or *anchylosed strabismus,* has been applied to this last condition. *Intermittent* strabismus is where the squinting comes on at certain regular or irregular periods, remaining each time for a longer or shorter period, and ultimately becoming, in most cases, permanent; it is commonly due to hypermetropia, and is easily curable by cutting the internal rectus—as the squint is usually convergent. Occasionally persons are met with, in whom strabismus manifests itself in one eye, and then in the other, and so on alternately—this is called *double alternative strabismus.*

Strabismus varies in its degrees, from a slight deviation of the optical axis, to one so extensive as to conceal as much as one-half or more of the cornea behind the nasal commissure (angle of the nose), that is, when it is a convergent squint. And, in nearly all instances, it will be found that vision is more or less defective in the affected eye, and, which, though sometimes due to some defect in the eye itself, is more commonly owing to its prolonged inaction, or perhaps to compression of the eyeball by the retraction of the muscles. In the last two instances, vision of the affected eye becomes much improved after an operation for the strabismus has been performed.

For the determination of strabismus, its cause, prognosis and treatment, I advise counsel, as heretofore in other affections, with a well-known, skilful and experienced oculist. Such an one will be found provided with opthalmoscope, prisms, strabometer, and all the required instruments for these purposes.

Where the cause is known, and can readily be removed, the strabismus will disappear; occasional cures have been effected by proper exercises of one or both eyes, with or without prisms; and where the squint is due to hypermetropia, the use of the proper convex glasses to neutralize the hypermetropia will often effect a cure of the strabismus at its commencement. But, in most cases of *active* strabismus an operation alone will effect the cure, in which certain muscles of the eye (the recti) are divided partially or completely, according to the degree and character of the strabismus. Of course, where the squint is due to paralysis, tumors or bridles, no benefit can be expected to follow an operation, until these difficulties are first removed, and then an operation may not be necessary. The operation for strabismus, although a highly valuable one, has been brought into much disrepute by being improperly performed by incompetent persons, or made in cases due to paralysis, tumors, &c., that could not be influenced by a mere sec-

6

tion of one of the muscles of the eyeball.    In these cases, the patient has, after the operation, found his strabismus fully as bad as before, if not worse.

Sometimes, it is impossible to overcome the strabismus at one operation, and two, or even three, operations at different periods may be necessary.    Again, an operation for convergent strabismus may be followed by an outward one, termed *divergent secondary* strabismus, which is the most awful disfigurement of the face imaginable; it is curable by the so-called "thread operation."

Many persons are troubled with what are called *motes, specks, spectral appearances*, as if flies, particles of dust, spider's web, gnats, butterflies, a cloud, filament, string of minute beads, or opaque round spots, &c., were before the eyes in the field of vision; sometimes but one eye is thus affected, at others, both.    These bodies are termed *muscæ volitantes* when they are movable, and *fixed muscæ* when they do not change their position.    There may be but one of them, or there may be many; they are of various sizes and forms, and often greatly interfere with vision.    The movable muscæ, however annoying they may be, are rarely, if ever, of a serious character, being usually due to some more or less opaque particles in the vitreous humor, or upon the cornea; they may be removed by hygienic measures, rest of the eyes, constitutional treatment if required, use of blue glasses, avoidance of any kind of excesses, and especially of bright lights, or of using the eyes by artificial light. Occasionally, these movable muscæ may be followed by the fixed variety, and especially in cases where the person continues to over exert his eyes, and pursues no measures for their relief.

Fixed muscæ (*scotomata*) are dark spots constantly

seated in some particular portion of the field of vision. They never move from this spot, although they have an apparent motion depending, however, upon the motion of the eyeball. They are usually dark or black, of various form and size, and if confined to only one eye, are more readily observed when the other eye is closed. They may be due to deposits of blood, lymph, or other substances in the vitreous humor, or to insensible spots on the retina, the result of some disease not yet known. Sometimes, the spots enlarge, interfere with vision, and terminate in total blindness; at other times, they may remain for many years, even for a life time, without occasioning any further disturbance to vision than that caused by their simple presence. Those who are affected with them are constantly annoyed by their presence whenever they attempt to read, write, sew, &c. The appearance of a dark fixed spot in the center of the field of vision among parties who are making investigations with the microscope, telescope, or other optical instruments requiring considerable illumination, and more particularly when this illumination is derived from artificial light, is an unfavorable indication, and should be promptly treated by the measures named below. If the use of the instruments be persisted in, complete blindness will ultimately be the result, more especially when the dark spot becomes more manifest and constant, or when it increases in size.

The first and most important point in the treatment, is absolute, positive rest to the eyes; the secondary points are, proper attentions to hygiene in every respect, and, should there be any disease of the liver, stomach, kidneys, heart, or brain, or any constitutional taint, the proper medicines and treatment to remove these maladies. However, fixed muscæ are rarely curable, but the above named course may prevent them, in many instances, from becoming of a more serious nature. As certain serious maladies of the eye, for instance, of the choroid and retina, have these fixed spots as one of their symptoms, a person troubled with them should at once apply to an oculist, and undergo a proper examination of his eyes, that he may use measures to overcome as promptly as possible, any existing disease of the deep seated parts of the eye.

# CHAPTER X.

## SPECTACLES.

BEFORE thoroughly entering into the subject of spectacles, their use, and abuse, and a discussion of the modes of adapting and wearing them, let us trace the history of this important aid. To do this, we must rely upon those whose researches offer us the nearest clue to their correct history. How many years it is since their first discovery, would be very hard for us to determine. We are told that in the recent excavations at the ruins of ancient Nineveh, not the least remarkable among the many interesting relics brought to light, was the discovery in the treasure house—a rock crystal lens. From this we may infer that the ancients were acquainted with lenses and their uses. How can we account for the perfection which they attained in cutting precious stones, if they did not use magnifiers? How could men attain such perfection in the other various branches of mathematics, mechanics, &c., &c., and yet leave this subject untouched which each drop of dew sparkling in the morning sunlight would suggest?

In 1575, MAUROBICUS, of Messinia, pointed out to the world what he considered the causes of myopia (short-sight), and presbyopia (long or far sight). Viewed in the light science has now thrown upon those conditions of the eye, his conclusions were erroneous—he was the first, however, to explain the benefits of concave and convex lenses to those conditions of the sight. Further back, however, we must go, if we would trace the first use of spectacles. M. CAESMAKER, a Belgian optician, who is said to have made diligent researches among the ancient records of the convents and monasteries, relates that ROGER BACON, who occupied the chair of philosophy at Oxford, passed some years at the convent of Cordeliers, in Lille, before he received that appointment. During his stay at Lille, he formed a friendship with a learned theologian, HENRI GOETHALS, known better as Doctor SOLEMNEL, and with the chronicler, PHILLIP MUSSCHE. To these he paid annual visits during the vacations at Oxford, and availed himself of the opportunities of procuring from the Belgian glass manufactories, fine glass fit for optical purposes which could not be obtained in England. With this glass, polished by himself, he made lenses, and communicated to his learned friends the secret of spectacles. Thus they became known in Flanders. Italian writers have attributed the invention of spectacles to the Dominican monks—but their knowledge was derived from this source. COOPER remarks that during the pontificate of MARTIN IV., who died in 1286, a question arose touching the interests of certain monks, who confided their defence to HENRI GOETHALS, he being sixty years of age, used spectacles, the glasses for which had been given him by his friend ROGER BACON. Arriving in Tuscany, GOETHALS visited his old friend NICOLAS MESSO, Prior of the Dominicans, and stayed among them two or three weeks; it was in this way that ALLISANDRO DI SPINA, the Dominican, became acquainted with the use and manufacture of spectacles.

As might be supposed, in those days a pair of spectacles was an inestimable treasure, whether made of iron, gold or silver, and carefully willed from one generation to another. We read that in the Inventory of CHARLES V., made after his death in 1558, there are enumerated among his valuables, twenty-seven pairs of spectacles!!

We often hear it remarked that spectacles are old-fashioned, and nose-glasses of late invention. On the contrary, until of late years, all lenses for improvement of vision, were mounted in nose-glass frames.

MANNI states that the art of making spectacles was carried to China by the Florentine Jesuits in 1697. COOPER, however, thinks that the use of spectacles in China, dates farther back than that. From one progressive state under the spur of the competition of the various manufacturers, and the universal demand for good, as well as cheap, articles, the art of making spectacles has been brought to the present high state of perfection. Now, instead of being considered as a blessing which might be left as an heir-loom, we find them in every household, as a common family article—however *common* they may be, we cannot detract from their value. The world is indebted to the pioneers in the discovery and manufacture of one of the greatest comforts of life, "for it is not too much to say that through the aid of spectacles we continue in the enjoyment, even in old age, of one of the most noble and valuable of our senses. They enable the mechanic to continue his labors, and the artist to display his skill in the evening of life, the scholar pursues his studies by their help, adding to the knowledge of others, and recreating his own mind with intellectual pleasures, thus passing days and years in satisfactory occupation that might otherwise have been

devoured by melancholy, or wasted in profitless idle-
ness." While thankful for this material aid of vision, it
should be a matter of careful consideration with every
purchaser, *what do I require? where shall I obtain?* We
can answer, purchase only of those who are thoroughly
versed on this subject, and you are safe. A hap-hazard
selection may prove injurious, instead of beneficial, to
vision.

## CHAPTER XI.

THE principal part of the *Spectacle* being the lenses, let us examine the different kinds and the principles of their action. Lenses are transparent bodies, possessing the power of increasing or diminishing the convergency of the rays that pass through them. Those lenses which are generally used in the manufacture of spectacles, are the only ones we shall mention in connection with this subject. They are six in number, viz:

FIG. 6.

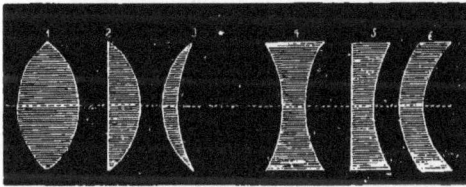

1. (Fig. 6.) A *double convex lens* is a solid disc of glass, having two spherical convex surfaces. When the convexity of each surface is exactly alike, the lens is termed *equally convex;* when they are not alike, one surface being

more convex than the other, the lens is termed *unequally convex.* The equally bi-convex lenses are those used in spectacles, and are termed by oculists *plus* or positive; thus, when a bi-convex glass, say No. 30, is ordered, it is written +30. It converges the rays of light.

2. A *plano-convex lens* has one of its surfaces plane, and the other convex. It is also a convergent lens.

3. A *meniscus* is a convexo-concave lens, one surface being concave, the other convex. As the convexity of a meniscus exceeds its concavity, the two surfaces would, if continued, meet each other, and this arrangement makes a convex lens. These lenses also form what are called *periscopic* glasses, to which an undue importance has been attached. (*See No. 6.*)

4. A *double concave lens* is a solid disc of glass, having two spherical concave surfaces. When the concavity of each surface is equal, the lens is said to be *equally concave;* when not equal, it is termed *unequally concave.* The equally double concave lens are those used in spectacles, and are termed by oculists *minus* or *negative;* thus, when a bi-concave glass is ordered, say No. 12, it is written —12. It diverges the rays of light.

5. A *plano-concave lens* has one of its surfaces plane, and the other concave. It is a divergent lens.

6. A *concavo-convex lens* has one surface concave, and the other convex; but as the concavity is greater than the convexity, the two surfaces, if continued, would not meet. This forms a concave or divergent lens, and has also been called *periscopic,* or *negative meniscus.* (*See No. 3.*)

The positive or convergent lenses have the effect of

increasing the visual angle,* and hence give an enlarged or magnified image of objects seen through them; the extent of magnifying power being dependent upon the degree of curvature of the convex surfaces as well as upon the degree of refractive power of the lens—the focal distance of the lens is also dependent upon these same conditions. The refractive power of transparent bodies varies, and that which belongs to one of these bodies is termed its *index of refraction;* thus, the index of refraction for the atmosphere, is 1,000294; for the diamond, 2,439; for sapphire, 1,794; for English flint glass, 1,830; for ordinary glass, from 1,525 to 1,534.

In addition to the above lenses there are also *cylindrical glasses,* which are exclusively employed in that abnormal condition of refraction in the eye, called astigmatism. These have the form of a segment of cylinder. Astigmatism, as heretofore remarked, is the result of an

FIG. 7.

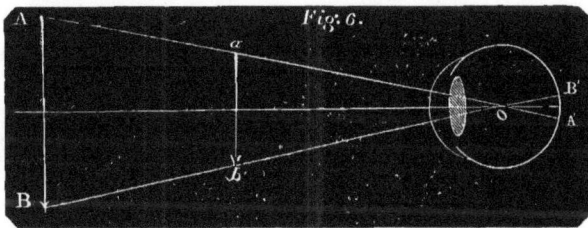

*By *visual angle* is meant the angle formed by two lines passing from the extremities of the object looked at to the optical center of the eye; the angle A o B, for instance in figure 7, is the visual angle under which the object AB is seen. Hence, it may readily be understood that the more distant the object is from the eye, the smaller will be the visual angle under which it is seen; and also that an object *ab* smaller than AB, but situated nearer the eye than the latter, may produce the same visual angle. On the other hand, on the side of the retina there is formed an angle A' o B' opposed by its summit to the visual angle; now, as the angles opposed by their summits are equal, we may exactly know this angle having its base resting upon the retina, by measuring the external visual angle AoB. We may thus, by the aid of quite simple calculations, determine what are any object whatever placed at a determined distance from the eye, subtends upon the retina. It is the measure of this angle that has been taken by SNELLEN for his new scale of test-types.

inequality of curvature in certain meridians of the cornea, more commonly in the vertical and horizontal diameters, in consequence of which, some objects are seen clearly and others confusedly, in accordance with the character of the astigmatism—and neither positive nor negative lenses alone will remedy the difficulty. The degree of the astigmatism is deduced from the difference of refraction in the two principal meridians.

To render vision clear, the refraction of these various diameters must be corrected by artificially increasing or lessening their curvature. A cylindrical lens produces this effect for only a single diameter, that is to say, it causes the rays of light to converge or diverge according to the perpendicular diameter of the axis of the lens, while it has no effect upon the luminous rays that pass parallel to the axis. There are three kinds of cylindrical lenses used: 1, *simple cylindrical glasses* which may be either convergent or divergent—the convergent or positive being bi-convex, plano-convex, and convexo-concave; and the negative or divergent being bi-concave, plano-concave, or concavo-convex. 2. *Bi-cylindrical glasses* that have two curved cylindrical surfaces, the axes of which are directed perpendicularly to each other. 3. *Sphero-cylindrical glasses*, one of the surfaces of which presents a spherical curvature, and the other a cylindrical curvature.

Even at the risk of some repetition, it may be useful to indicate here the forms of astigmatism in which these three principal types of cylindrical lenses are used. Thus—

1. In *simple myopic astigmatism*, which is characterized by a certain degree of *myopia* in only one meridian, and *emmetropia* in the other, the difficulty is corrected by a

*simple negative cylindrical,* of a focal distance corresponding to the degree of the astigmatism.

2. In *compound myopic astigmatism,* characterized by a
certain degree of *myopia* in both meridians, we remedy
it by a *negative sphero-cylindrical* lens.

3. In *simple hypermetropic astigmatism,* characterized
by *hypermetropia* in only one meridian, the other being
*emmetropic,* a *simple positive cylindrical lens* is required.

4. In *compound hypermetropic astigmatism,* characterized by *hypermetropia* in both meridians, a *positive sphero-
cylindrical lens* is required.

5. In *mixed astigmatism,* characterized by *myopia* in
one meridian, and *hypermetropia* in the other, a *bi-cylindrical lens* is required.

As to the principles of the action of convex and concave glasses, the following demonstration from COOPER,
will fully illustrate them:

FIG. 8.

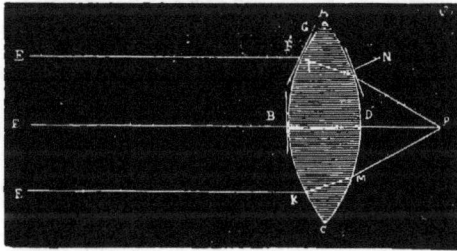

Let A B C D, Fig. 8, be a transverse section of a double
convex lens and E F, E B, E K, parallel rays of light falling upon such lens. As the ray E B falls at right angles
to the tangent of the arc, A B, it passes through the lens
in the direction B D, in a straight line with E B. But
the ray, E F, on passing into the lens, that is from a *less*

to a more refracting medium, and in a direction not at
right angles to the tangent, F G, is refracted towards
the line, I F, which is perpendicular to the surface of the
lens at the point at which the ray, E F, entered. Having
traversed the lens, it undergoes further refraction on
leaving it; but in this case, because it is leaving the
more refractive, and entering the less refractive
medium, it is refracted from the perpendicular, N N,
in the direction, N P, and meets the axis at P, which is
the focus. If the lens be of glass, and equi-convex, the
distance of the focus behind the lens will be equal to the
radius of either of its surfaces; the effect, therefore, of convex lenses is to render parallel rays convergent.

The action of concave lenses upon light is precisely
the converse of the action of convex lenses, rendering
the rays divergent.

Fig. 9.

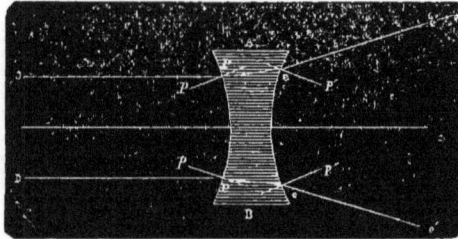

A B, is a transverse section of a double concave lens;
c and d are parallel rays incident upon it. The rays
on entering the lens are refracted toward pp, the perpendiculars to the surface at the points at which the
rays, C and D, entered, and on leaving the lens they
undergo a further refraction; in this case, from p'p' the
perpendiculars to the surface at the points at which they

emerge, that is, they will diverge in the direction
e e'ee'.

The kinds of glass used in optics, are as follows: *Flint-glass*, or crystal, which is a silicate of potassa and lead;
*crown glass*, which is a silicate of potassa and lime; and
*rock crystal*. *Flint glass* should be rejected, because it
decomposes the light too much and scratches easily.
*Crown glass* is preferable, and its composition should be
about the following: white sand 120, carbonate of
potassa 35, carbonate of soda 20, chalk 20, arsenious
acid 1. As to *rock crystal* or quartz, it ought to be
rejected on account of the double refraction it possesses.
It can only be used by cutting it perpendicular to the
axis of the crystal. *Rock crystal* is more refracting than
*pure crown glass;* from whence it follows that to obtain
a given focus with it, the curvatures must not be so
great as with crown glass. All glasses, whether of ordinary glass or rock crystal, are made in a similar manner; but the numbers of rock crystal lenses differ from
those made of crown glass, as to focal length and magnifying power.

Good spectacle lenses should have an equal polish;
clear, regular borders; freedom from all scratches, bub-
bles, or other blemishes, and should be very clear.
Placed upon a piece of white paper, they should present
no tint, and objects seen through them should appear
clear and distinct. It is, indeed, highly important that
they be well ground and polished, for every variation
from the true curvature will interfere with the refraction, and the image formed will necessarily be ill-defined
and blurred.

The lenses for spectacles, and, indeed, all lenses of
good quality, are manufactured in Europe. The English furnish the world with both the *cheapest* and the

most expensive lenses. The finest lenses are OSBORNE's first quality convex lenses; they are unequaled for their purity and hardness, and for practical use they are preferable to genuine pebbles, for which they are often sold. The English lenses are usually made smaller than the French, and can only be used in English frames, hence, because of their size, the first quality of French and German are preferred by opticians, as being capable of adaptation to any frame.

The Rathenow (German) glasses have acquired quite a reputation in certain parts of the United States, but after a thorough examination of them, I find them no better than the first qualities of French lenses.

Convex lenses for spectacles are numbered according to their focal distance, which is always expressed in inches; thus, No. 60 has a focal length of 60 inches; No. 30, of 30 inches; No. 8, of 8 inches, and so on. The weakest glass, or the one having the longest focal distance being numbered the highest, and *vice versa*.

Spectacle lenses may be classified into the following grade of numbers:

| 60 inch.............. | .............14 inch. |
|---|---|
| 48 " .............. | .............13 " |
| 42 " .............. | .............12 " |
| 36 " .............. | .............11 " |
| 20 " .............. | .............10 " |
| 24 " .............. | ............. 9 " |
| 20 " .............. | ............. 8 " |
| 18 " .............. | ............. 7 " |
| 16 " .............. | ............. 6 " |
| 15 " .............. | ............. 5 " |

Lenses, the focal distance of which measures *less* than 5 inches, are termed cataract lenses, and are used in cases where an operation for the removal of cataract has been performed.* The range of cataract glasses are—

| | | |
|---|---|---|
| 5 inch................ | ...................... | |
| 4¾ " ................ | ....................2½ inch. | |
| 4½ " ................ | ....................2¼ " | |
| 4¼ " ................ | ...................... | |
| 4 " ................ | ....................2 " | |
| 3¾ " ................ | ....................1¾ " | |
| 3½ " ................ | ....................1½ " | |
| 3¼ " ................ | ....................1¼ " | |
| 3 " ................ | ...................... | |
| 2¾ " ................ | ....................1 " | |

Very rarely a lower lens than 1¾ inch is used, and generally, less than two inches are not to be found in the packages of one assorted dozen, each, met with in the wholesale stores.

Concave glasses are numbered in two ways. 1st. *The French and German* number the concave glasses according to their focal length (which is negative), or in other words, the weaker the focus of the glasses, or rather the longer their focal distance, the higher are they numbered (the same as in convex or positive lenses). And the stronger the focus, or the shorter their focal length, the lower are they numbered; by this method, the same number of a concave and convex lens placed in apposition, neutralize each other: thus, a number 12 concave glass placed against a number 12 convex, will neutralize the effect of each other, and objects will appear the same as if seen through a plate of plain glass. 2d. *The English*, with the exception of the Royal London Ophthalmic Hospital at Moorfields, pursue an entirely opposite

---

* See Chapter VIII.

method, numbering the stronger focal glasses with the high numbers, and the weaker ones with lower numbers, which is very inconvenient. In this country, the French and German plan is almost universally adopted on account of its simplicity, and the readiness with which the number or focal length of any concave lens may be determined, by merely finding the convex lens that will neutralize its divergency, and make objects appear the same as if looked at by the normal naked eye.

The comparison of French and English numbers are as follows:

*Concave English* about *equal to French*

No.   1......................to...................36
"    2...................... "....................27
"    3...................... "....................24
"    4...................... "....................20
"    5...................... "....................15
"    6...................... "....................12
"    7...................... "....................11
"    8...................... "...................... 8
"    9...................... "...................... 7
"   10...................... "...................... $6\frac{1}{2}$
"   11...................... "...................... 6
"   12...................... "...................... 5
"   13...................... "...................... 4
"   14...................... "...................... $3\frac{1}{2}$
"   15...................... "...................... $2\frac{3}{4}$
"   16...................... "...................... $2\frac{1}{2}$
"   17...................... "...................... $2\frac{1}{4}$
"   18...................... "...................... 2

*Pebbles*, which have been so highly recommended by parties terming themselves opticians, but who are, in

the majority of cases, only pedlers of various grades,
deserve a moment's notice. The qualities of pebbles are:
purity, extreme hardness, and not being so easily
scratched, in use, as the common lenses. The qualities
attributed to them are miraculous. "They will renew
the sight, they will cool the eye, and they will remove
any possibility of the wearer's sight deteriorating."
Such results are offered by strolling adventurers under
the name of opticians, to those who will buy these spectacles. It is useless to say the whole system of argument is a falsehood, from beginning to end. "Yes, my
dear sir," it may be answered, "but I never found such
spectacles in my life as I bought of Mons. M——, who
visited us some years since; they fitted me perfectly, and
the glass was so good." "Exactly," I answer, "but the
reason is simply that prior to that, you had depended
entirely on jewellers and pedlers for your spectacles,
and not upon reliable opticians. Mons. M——'s spectacles suit you so much better than any you have ever
had, not on account of any particular perfection in what
he termed pebbles, but in *his ability to fit you.*" Admitting, then, that pebbles are particularly hard, and not
easily scratched, it becomes of little consequence when
we consider that by the *gradual,* but sure, deterioration
of sight, new lenses of greater power must be substituted for the old ones, at least once in every two or three
years, and the old glasses being of no further value,
must be thrown away. The pure English glass will not,
in two years time, scratch sufficiently to obstruct vision
in the least, and being quite as transparent as the pebble,
they are just as useful and greatly less expensive. It is,
therefore, very foolish and extravagant to pay five or
six times as much for pebbles, when pure English glass
will answer equally well, and in nine cases out of ten,

the probabilities are that those who pay for pebbles, get only English or French glasses, as even opticians themselves often find it difficult to distinguish the fine OSBORNE glass lenses from the pebbles.

Thus far, I have only referred to clear white lenses; but there are others designed to protect the eyes from too intense a light, etc. These are glasses of various colors, as green, blue, and blue black, smoke, or neutral tint. Colored *plain* glass is chiefly used to protect the eyes from the irritation produced by the rays of bright light, especially when these organs are weak, inflamed, or laboring under an undue sensibility to light, &c.; they also protect from wind, dust, and other irritating agents. For many years, the Chinese have used "a substance called 'cha the,' or tea stone, from the resemblance of its transparent hue to a weak infusion of black tea, for the purpose of subduing the glare from the sun. It is, probably, a smoky quartz or silex allied to the Cairgom of Scotland."

Colored convergent or divergent glasses are used in cases of short-sight or long-sight when associated with weakness, irritation, &c., and will be found very useful among those who use their eyes very much at night by artificial light: they act by diminishing the intensity of the light, and thus relieve the eyes from a great deal of fatigue and wearisome effort.

*Green* glass alters the tint of objects looked at, and is frequently annoying to the eye; it should never be worn. The idea so prevalent in former years, and even now upheld by ignorant persons, that great virtue exists in green glasses, has long since become a matter of simple amusement among the better informed. No virtue, whatever, exists in color, save as protectors to the eye, under certain circumstances. Yet, from the fact that

many persons, not even excepting some physicians who
ought to know better, attribute some unknown virtue to
green glasses, there are persons who will refuse any
other color, and hence opticians are obliged to keep a
full supply of green spectacles on hand.

It may be said by some, "why, my family physician
recommends me to wear green glasses." "Yes, my dear
friend, that may be so, but how much does your physician know about the eye and its requirements? Certainly, nothing whatever, in a practical view, if he gives
you such advice." This habit of receiving advice from
uninformed parties regarding the wearing of spectacles,
is one which oculists and opticians find great difficulty
to counteract.

JOHN QUINCY ADAMS being asked on one occasion,
his opinion of a certain subject, replied that he could not
give an opinion upon a subject about which he knew
nothing. And if the ignorant advisers of those requiring aids to vision, would be as guarded in their opinions
and advice as this great statesman, a great many eyes
would be saved a vast amount of labor, fatigue and
suffering. In sickness, do we ask some friend to point
out from which bottle in the apothecary's store we shall
take our medicine; or, do we not, relying upon the
better knowledge of the apothecary as regards his medicines, allow him to make the required selection for us?
Let every one, then, unacquainted with the eye and its
requirements, act upon the same principle with the
oculist or skilful optician, who is able to suit the eyes
vastly better than could possibly be done by those who
have no knowledge whatever upon the subject.

*Blue* glasses, although they alter the tint of objects,
yet do so in a less degree than green, and are decidedly
much better for ordinary use. The same as with green

glasses, they furnish the sensation of the complementary colors when the light strikes upon the sides instead of passing directly through them; with blue glasses these rays then appear yellow. The irritating rays of light are those termed "chemical," and blue glasses lessen the amount of the rays very considerably. WECKER states that, "for the removal of dazzlings caused by too intense a light, blue glasses are the most proper, as they do not plunge one into a state of semi-obscurity, as do the neutral tint or London smoke glasses which eliminate indifferently a variable quantity of all the rays of the solar spectrum." Yet notwithstanding this, many cases will be found where blue is intolerable to the eyes, and where the neutral glasses will prove far more agreeable to these organs.

*Neutral, smoke,* or *black-blue glasses, (London smoke),* do not change the tint of objects, they only show them less colored, or rather with a darkened light, and, unlike the blue and green, do not give rise to the complementary colors. These neutral glasses appear to be preferred by persons generally who require colored ones, on account of the pleasant sensation they produce, their effect being similar to that caused by a cloud passing before the sun. In the case of the Chinese, heretofore referred to, using the "tea-stone," they appear to have manifested considerable judgment. I may state here that the inhabitants of the polar regions protect their eyes from the effects of the reflection of the sun on the snow, by placing small capsules before their eyes, pierced with a minute hole just opposite the pupil of each eye.

I have been shown a specimen of cataract lens, termed "Isochromatic." They are a combination of two plano-convex lenses, with a very thin slip of plain blue or smoked glass cemented between them. The effect is to give the lens an equality of color which we do not find in the usual cobalt blue or smoke bi-convex lenses. Upon a trial of them, however, I found one very great objection, that would render them of little practical utility, viz.: in the rough, every day usage to which spectacles are subjected, the balsam or cement cracked, and thereby effectually interfered with the rays of light passing through them, and, consequently, the spectacles were of no further use. Such lenses may be of value to the oculist in making ophthalmoscopic examinations of the eye, as recommended by R. B. CARTER in his Translation of ZANDER on the Ophthalmoscope, p. 77. Plano-concave lenses formed into bi-concave lenses by a similar method, have been recommended by Dr. J. Z. LAURENCE, to prevent the peculiar dazzling appearance conferred on objects by very deep concave glasses.

Lenses are used for benefitting vision, or diminishing fatigue of the eye, &c., under the form of *spectacles*, and *nose* or *eye glasses;* but, for continued use, the latter form is objectionable, as they cannot be kept adapted properly in front of the eye for any great length of time. These two forms consist of the lenses and their mounting or frame. The mounting consists of gold, silver, shell, horn, hard rubber, or steel, and is manufactured into various shapes. The light steel frames are generally preferred. In the adaptation of spectacles to parties requiring them, great care is necessary that the bridge of the frame fits exactly over the curvature of the nose, that it is of such a length as to have the center of the lenses correspond with the pupil or the optical axis, and that it is so curved from before backwards as to bring the lenses directly in front of and very close to the eye, but not so as to touch the eyelashes, with a slight inclination backward, from above downward.

Spectacles are made in almost an infinite number of models—to specify them would be an almost endless task. I shall, however, endeavor to place before the reader *cuts* of the most prominent models in use at the present day; recommending such as are easily worn, and which, from their positions upon the face, afford the most distinct vision. It should be remembered that *spectacles*, like clothing, have their fashionable and unfashionable styles. In the East, the oblong and octagon pattern is the only one worn. In the West and South, none of that style can be sold, purchasers preferring the oval and pantascopic styles.

Besides the various glasses already referred to, there are some others to which I will make a brief reference. *Prisms* are glasses cut with perfectly plane surfaces, but inclined upon each other at angles varying from 1 to 20 degrees. The luminous rays passing through them are refracted towards the base of the prism; or, in other words, objects seen through a prism appear deviated towards its apex. In consequence of this, if, while looking at an object with both eyes, one of these prisms be placed before an eye, the object will appear double. But if the degree of inclination of the two planes is not great, we may be able to fuse the two images into one. Prismatic spectacles have been used to correct certain cases of diplopia or double vision, and strabismus or squint.

*Stenopaic Spectacles* are formed of a black metallic plate, pierced in the center with an orifice about four-fifths of an inch in diameter, upon which is fixed a movable plate carrying a small conical tube, the base opening on the side towards the pupil. By means of a particular mechanism, the tube may be changed in its position and be placed directly in front of the cornea or crystalline, that the luminous rays may penetrate the eye. This is designed to overcome the confusion of vision resulting from the vicious refraction of luminous rays upon partial opacities of the cornea, or of the crystalline. A similar spectacle, but without the movable plate, is used in mydriasis, or morbid dilatation of the pupil, and in cases of absence of the iris.

# CHAPTER XII.

## SPECTACLES—CONTINUED.

ALL spectacles should be adapted to the contour of the face of the wearer. To obtain distinct vision, it is *absolutely necessary* that the optical axes of the eye and of the lens should coincide, no matter what may be the defect which the glasses are designed to remedy. This is a point which should always be observed both by the optician and the purchaser, because it is essentially necessary to distinct vision, yet, generally, little attention is paid to the matter, the necessity being little understood or appreciated. Many even do not regard it as very essential that the *frame* should fit the face as well as the lenses should suit the condition of the eye. *One is just as necessary to perfect vision as the other.* It will not do for a large faced man to put on a narrow, or small frame, for when he does, the lenses, instead of being thrown directly into such a position that he looks through their centers, are so placed before the eye that he looks through the outer edges of the glass, and thereby loses the full benefit of the lens—the image formed on the retina being imperfect, instead of clear and distinct. The effect of a large pair of spectacles on a small weazen-faced person, is of a similar character; he will see through the outer edges of the lenses—and as the optical axes of the eyes and of the lenses will not coincide, the result will be the same as in the opposite case above referred to.

The distance between the centers of the pupils, and that between the centers of the glasses, should be equal; for, as I have before observed, the clearest vision is obtained when the wearer looks directly through the centers of the lenses—hence, the eyes will have a constant tendency to look as nearly as possible through the axes of the glasses. "If the lenses be too far apart, the eyes in striving to accommodate themselves, will acquire a tendency to an out-squint, while if the glasses are too near together, there will be, for a similar reason, a tendency to an in-squint. The frames should not only correctly adjust the glasses, but should maintain them firmly and steadily before the eyes." Great care should be taken both by the purchaser and the optician, that the spectacle frame be perfectly straight, and the glasses be set in the same plane—for a change in the position of *one* of the glasses of a pair of spectacles on the face of an individual, will materially and injuriously change the effect of the two glasses upon the sight. To illustrate, let us examine

Fig. 10.

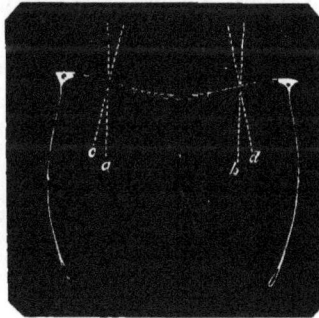

We know that the axes of the glasses, c, d, should coincide with the axes of vision, a, b. But in Fig. 10 they do not, for the ocular axes (of vision), a, b, are altered in their direction and become convergent; the glasses ceasing to be in the same plane, their axes lose their parallelism. Hence, *iron* spectacles, on account of their being so easily bent or put out of shape, should be discarded for those made of a firmer metal. And, as is often the case, when spectacles do get out of shape, there should be no delay in having them re-put in perfect order before wearing—unless, indeed, the wearer is indifferent to the preservation of his sight.

The best and most distinct vision is obtained by wearing the double convex or concave spectacles, according to the hypermetropic, presbyopic, or myopic condition of the eye. Dr. WOLLASTON was the first to introduce the lens which he termed "periscopic," because it enabled the wearer to see the various objects anteriorly around him without moving his head, being ground hollow on the side next to the eye and convex on the other, somewhat in the shape of the waning moon; the degree of the focus for myopic or presbyopic eyes being effected by the degree of concavity and convexity given to the glasses. This is a good glass when used only for walking about, or for out-door work, as they enable the wearer to see distinctly at all points around him; but in reading or writing, they do not form a distinct image upon the retina, neither do they form a well-defined and sharp image upon the wall.* On the contrary, double-

---

* The method of measuring a convex glass is to hold it near the wall or side of a room facing a window, or opposite the sun or lamp, and moving the lens slowly backwards and forwards until the image (formed by the lens upon the wall or side of the room) of either the sun, the lamp, or of any object situated between the sun and the lens, becomes the smallest and most distinct; the distance in inches between the lens and the wall, at that moment, is the focal length of the glass.

convex lenses give well-defined images both upon the retina and the wall.

The position of the spectacles upon the face should not be the same in all cases. In *myopia*, the object should be to have the spectacles so adapted to the face, that the eyes, in looking *forward*, shall have their optical axes (or their pupils) to correspond with the centers of the lenses. In *hypermetropia*, the same rule will apply. In *presbyopia*, the object should be to so adapt the spectacles that the wearer shall see directly through the centers of the glasses when he looks down upon the work, sewing, reading, or writing, &c., which is before him; but in those cases frequently met with among aged persons, in which there is also a failure for *distant* as well as near vision, the individual will be required to wear two pairs of spectacles—one, adapted, as already stated, for reading, writing, &c.; and the other, worn in a position similar to that named for myopia, for distant vision, promenading, &c.

If the spectacles for the presbyope are placed directly before the eye, the head must be inclined while reading, &c. But if they be so fitted as to incline to a certain degree below the eye, the position of the head need not be altered any more with than without spectacles. The best spectacles, therefore, for presbyopes are those which incline the lenses downwards, so that their axes shall be perpendicular, or nearly so, to the plane of the book or object looked at. This inclination should be such that the axes of the lenses form an angle of about 30°, with a horizontal line drawn from the eyes during distant vision. And as the eyes always converge, more or less, in near vision, a convergence should also be given to the lenses, so that the optical axes shall form continued straight lines with the axes of their corresponding lenses. This convergence, however, is rarely attended to in the fitting of spectacles, though of great importance to vision. In

FIG. 11,

We have a representation of a spectacle which is generally termed *Pantascopic.* I consider it by far the best model of a frame for the far-sighted to wear. At a glance, it will be seen that the arms or branches of the frame, instead of being exactly at right angles with the glasses of the spectacle, are slightly inclined, as already referred to. The effect of this slant is to throw the lenses obliquely, under or before the eye, so that upon casting the eyes downwards, upon a book for instance, the head being held somewhat erect, the optic axes and the axes of the glasses coincide with each other. The bridge, arch, or nose piece is placed higher up above the upper border of the lens holders than is common to ordinary spectacle frames; and the oval fenestræ or lens holders have their upper border somewhat straightened, so as to partially cut off the upper segment of the oval usually formed in ordinary spectacle frames by this border. These dispositions of the frame allow the lenses to fall somewhat below the range of vision when accommodated for distant objects, the wearer being able to look over these borders; while for writing, reading, &c., the spectacle is ready for use at all times. These frames may readily be worn while walking, or while working on distant objects, since the glasses do not interfere with distant range, which, in the common styles of spectacle frame, is very annoying.

Fig. 12.

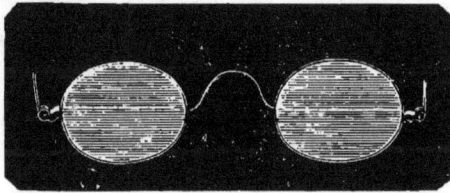

In the above (Fig. 12) we see the style of spectacles now generally sold for a reading spectacle; the bridge or nose piece is styled "English," to distinguish it from the K, or French, which is made in a style similar to the one represented in

Fig. 13.

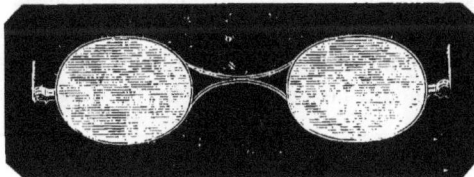

But the axes of the glasses are generally thrown too much out of the direction of the optical axes, to afford correct vision for presbyopes. While exactly the opposite may be said of them, as spectacles for the myopic or short-sighted, for whom they are more appropriate, as the glasses are brought before the eye and close to it, and, when in a state of rest,* the centers of the glasses and of the eyes coincide, and the wearer seems to be looking through one glass instead of two, which should be the case.

In Fig. 14 we have the style in which the spectacles of olden time were made.

FIG. 14.

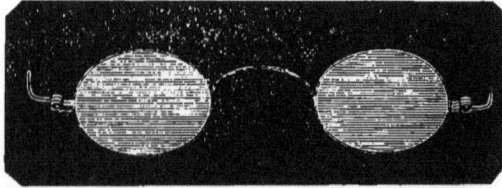

The pattern is now obsolete, however, excepting for very common "Yankee" spectacles made of German silver (with a single screw to act as a pivot on which the arm turns, and which also serves as a means of holding in the glass. Spectacles are generally made with a rivet on each side of the lens holder, to hold the arms in their places, and also a screw on each side to fasten the glasses in the frame). The glasses in these Yankee spectacles, are of the very commonest kinds. The French, also, make a similar spectacle of iron. These spectacles should never be worn by those who have any

---

* Accommodated for distant objects.

regard for their sight. Many persons wear these very cheap spectacles merely because they *are cheap*, arguing, that because they lose their spectacles every week or two, or once in every two months, cheap glasses will answer very well. Better, however, pay thrice the cost for a good article, even every month during the whole year, than to gradually have vision permanently impaired, or prematurely destroyed.

It is a very poor policy in buying glasses, to purchase a poor article simply because it is cheap, for the eye is too sensitive and too important an organ to be trifled with, or to endure a continual abuse without certain, and often speedy, deterioration of sight. For those whose myopia is of that character requiring them to wear spectacles continually, the best and most reliable

Fig. 15.

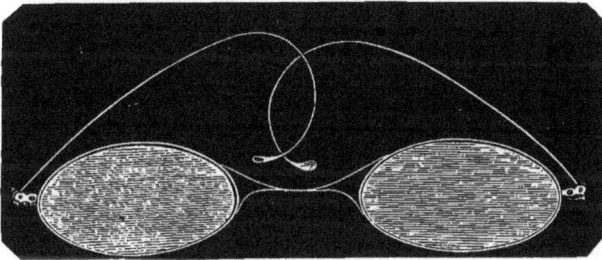

style is the above (Fig. 15), representing a crochet or riding spectacle. The frame is grooved into the glass, and, being very light and delicate, is almost invisible; the nose piece, as well as every part of the frame, is made of the best tempered steel, and when the arms are hooked behind the ears, the glasses are held firmly to their places before the eyes; the frame is hardly felt on

8

the face, and can not be readily dislodged from its position. The glasses in these frames are generally fine— but care should always be taken to purchase a *good* article, as the poorer qualities of crochet are worthless, and will last for a short time only.

FIG. 16.

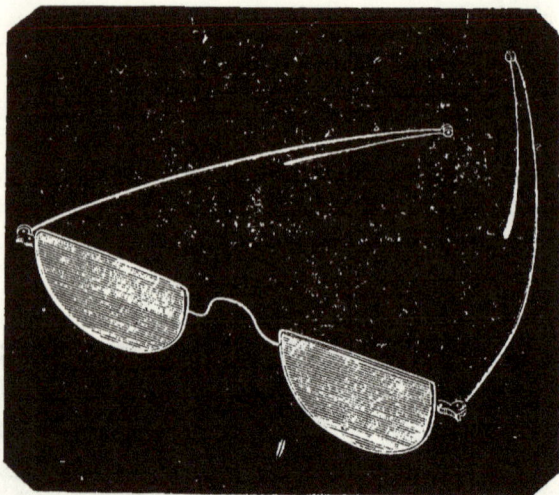

The above (Fig. 16) is the half eye, or, as formerly termed, "The Pulpit Spectacles." These are admirable for many purposes. But as the Pantascopic spectacles (see Fig. 11) possess all the advantages without the ill-shape, and want of symmetry of these, which form the great objection to them, they are generally preferred.

The double focus, or, as they are sometimes termed, "Split Spectacles," are made as in Fig. 17. The lenses in the upper halves of the spectacle are of a weak focus for viewing distant objects; the lower halves have a stronger focus for reading, writing, &c. FRANKLIN

was the first to invent this method of uniting two spectacles in one:—"Recently, it has been attempted to
attain the same object by grinding in the upper part of
the spectacle glass, the surface turned from the eye, with

Fig. 17.

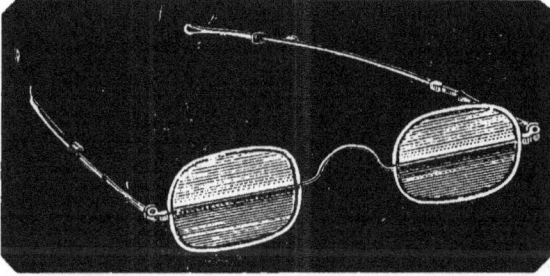

another radius. The glasses are prepared in Paris
under the name of *verres a foyer.*

Fig. 18, *a*, represents a superficial view of such a
glass (slightly positive above and strongly positive

Fig. 18.

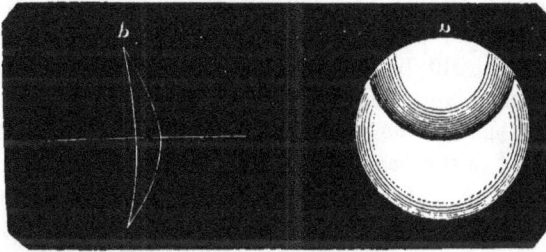

below); *b*, exhibits a section of the same." But this
style of glass, if not ground with great care, will be
found to act as a prism at the dividing of the two foci—
they are not generally available.

The objection to the split glasses is that the line which divides the two foci (Fig. 17) is very disagreeable and annoying to the wearer, except where by constant and continued use the eye becomes accustomed to looking *over* or under this line (as the case may be). They are convenient, it is true, after the eye has become adapted to them; yet they should be discarded, as they occasion great fatigue to the eye while accustoming itself to their use. In connection with this, I must again repeat what has been already stated, viz: *The eye should never be adapted by use to the spectacle, but, invariably, the glasses to the condition of the eye.*

In a former chapter it has been already mentioned that the eye is often abused and injured by exposure to intense lights. Spectacles are now manufactured to protect the eye when under such exposure to intense light, also for the prevention, as well as the alleviation of inflammation; and they have been brought to a high state of perfection. From the olden times when the Greenlanders wore their plate of bone before the eyes, that by it they might be protected from snow-blindness—or, from "The Mugusian snow shades," which were opaque plates, in which were cut only a narrow horizontal slit before each eye; some being of beaten polished silver, but most of yellow birch bark of the "same shape, bound at the edges with leather, and then fastened to the ears with thongs"—or, from the days when "the Yakuts adopted, for protecting the eyes, a very neatly made narrow meshed net of black horse hair, about six inches long, and broad enough to cover the eyes, having its elliptical border sewed to a thin piece of leather, in such a way that the side turned to the face becomes slightly concave; this was fastened with thin leather loops to the ears,"—we have arrived at the present standard of excellence in these various contrivances for protecting the eyes. Eye protectors may be divided into—

I. GOGGLES.

II. SHADES.

III. COLORED SPECTACLES.

Goggles are mainly used to protect the eye from contact with particles of dust, mites of cinder, &c., while traveling to and fro; also, to protect it from becoming injured by chaff and other particles while the farmer is threshing, &c. They consist of two fine wire-gauze cups or receptacles, one for each eye, and so arranged as to be adapted over the orbits, and to cover the eyes, being retained in place by an elastic band which passes around the back part of the head. In the front part of these wire-gauze cups, glasses are fitted, being either plain or tinted, (smoke, blue, green, &c.). (*See Fig. 19.*)

FIG. 19.

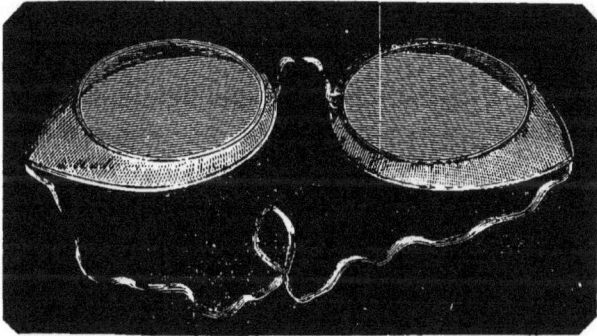

These goggles (Fig. 19) are made in a great many styles, and at almost any number of prices. The cheapest in use, as in threshing, costs only 50 cents, while the best velvet mounted around the edges so as to fit comfortably, cost three or four times as much. This is an article well known, and thoroughly appreciated by all travelers who have once used them, especially in railroad traveling.

FIG. 20.

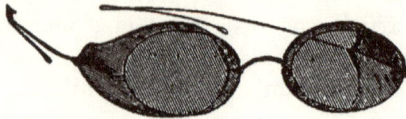

Fig. 20 represents an old style of eye protector, the receptacles consisting entirely of wire-gauze attached to their frames, without any glasses; these were used only for protection from dust, grit, and other solid particles. They were not designed for wearing in a room at night, nor for reading, writing, &c., prismatic colors being formed by the action of the rays of light upon the wire-gauze. A similar style of frame is now made and in use, plain glasses in front of the eye being substituted, however, for the wire-gauze.

FIG. 21.

An ingenious goggle (Fig. 21) was devised by COOPER for Sir EDWARD BELCHER's party before starting on their expedition to the polar regions in search of Sir

JOHN FRANKLIN; the material was papier mache, varnished on the outside, and dead blackened within, to prevent reflection. The eye pieces were edged with velvet to afford a soft cushion, and to completely exclude all side drafts. The slits were sufficiently large to permit vision, and the distance at which they were placed from the eyes was intended to protect the eyes from the steel, while the sides protected them from the glare by shutting out all unnecessary light.

Shades of every size are used to protect the eyes from sunlight; the small single shade for one eye being used only in cases of bruises, contusions, &c.

FIG. 22.

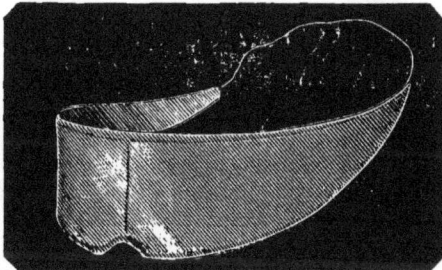

Fig. 22 represents a shade used by many for relieving the eyes from the glare caused by too great a light projected on an object looked at, as well as to protect the eyes from too intense direct light upon these organs themselves; it consists of a wire frame covered with green or black silk. I have used one for quite a considerable time, and find that in writing, with the light streaming from above, they are very useful and convenient, affording much relief to the eye, and preventing it from becoming readily fatigued or irritated.

Spectacles for the protection of the eye from annoying rays of light are made in the same styles as we have
already enumerated, colored or tinted plain glasses
being substituted, however, for the convex or concave
lenses. In

Fig. 23,

We have represented the four glass spectacles, opaque
or colored glass sides being attached to the lens-holder
of the spectacle, thus enabling one to exclude all side
rays, as well as those directly in front of the eye. They
were formerly made with convex or concave glasses
fitted in the front of the frame, and, by simply shutting
down the colored plain glass sides, a tinted lens was at
once formed. Of course, in our days, we do not require
such methods for attaining this end. An old gentleman
of this city resorted to these four glass spectacles to
obtain a double focus spectacle; thus, in the front frame
he placed the lens which suited him for distance, and in
the side frame a lens of such power as, when combined
with the front lens, would suit him for reading, &c. By
this method, when he desired his spectacles for distant

vision. it only became necessary to remove the side
glasses from the front ones. But such spectacles are
too heavy, and very few would find them at all desi-
rable. For shading the eye from irritating rays, they
have almost entirely gone out of use, being supplanted
by the "Coquille" or "shell spectacles," (*verres a bom-
bees.*)

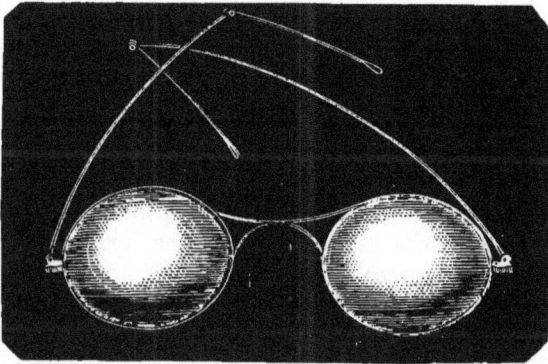

FIG. 24.

These glasses (Fig. 24) are tinted, and made somewhat in the shape of a watch crystal, being large and
bulging, the concave surface being next the eye, and the
convex or bulging surface being placed externally. The
eye is entirely covered, and yet there is perfect venti-
lation. During the late rebellion, these Coquille glasses
became very popular with the soldiery during their
marches in the sun. They are in great demand among
oculists, for patients suffering from excessive retinal sensibility, inflammatory affections of the eyes, &c.

Other spectacles are occasionally used, and to which,
in a work like the present, a mere reference is all that
is necessary. Thus, for patients affected with partial

opacities of the cornea or crystalline, DONDERS' steno-paic spectacles are often employed. The glasses are replaced by a black metallic plate, in which is pierced an orifice about 7-10ths of an inch in diameter; upon this is attached a movable plate carrying a small conical tube, its base opening at one side of the pupil. By means of a particular mechanism this tube may be changed in its position, and be placed in front of that part of the cornea or crystalline which permits the luminous rays to enter—thus preventing the confusion resulting from the vicious refraction of luminous rays upon the semi-opaque parts of the cornea or crystalline. For persons affected with mydriasis, the black metallic plate alone is used, being pierced in the center with a small hole, or with a crucial slit.

Spectacles with prismatic lenses (or glasses with per-fectly plane surfaces, but inclined one upon the other from 1 to 20 degrees,) have been used in double vision, strabismus, weakness of the recti muscles of the eyes, &c.; but they require to be worn, and the eye to be exercised with them, for a long time, in order to derive any bene-fit from them; hence, oculists prefer certain operations, as being more prompt, more efficient, and more perma-nent, in performing cures of these several abnormal con-ditions.

# CHAPTER XIII.

## NOSE GLASSES.

HAND or Nose glasses have of late years become very popular, owing to three reasons, viz:

I. The desire to conceal from friends and acquaintances an imperfection denoting the approach of old age. Those who by reason of the natural deterioration of vision are obliged to wear glasses, generally prefer to commence with nose glasses, simply because they dislike to be seen wearing spectacles, as they imply age and its infirmities. This is but a natural tendency; it is inherent in the human heart to resort to all sorts of artifices to conceal a natural and unavoidable defect, and certainly there can be no fault found with the perfectly justifiable desire to improve our appearance, and make ourselves as pleasing as possible in the sight of our fellow creatures—a love of the beautiful, of order, of symmetry, prompts this desire. But, we have no sympathy with those who, for the purpose of attracting notice or obtaining a kind of notoriety, assume defects not natural to them, and which brings us to reason—

II. Nose glasses are sometimes worn by simple minded persons who do not naturally require them, solely to satisfy vanity and egotism. The fop is incom-

plete without his eye glass.* His style would be imperfect without this attraction; and thousands of nose
glasses are sold to ladies and gentlemen who use them
merely as toys by which to attract attention to themselves.
Of course, this class of persons will be found princi-
pally in our larger cities, and they form a source of no
small income to the opticians of London, Paris, New
York, &c., &c.

III. Their utility and convenience for momentary
use. The "pince-nez," or nose clips are very convenient, owing to their being so easily opened and brought
into position at once; but while so serviceable to the
merchant, lawyer, &c., for merely occasional use, they
are decidedly injurious when used continually in lieu of
spectacles—because, it is very seldom they are placed
in exactly the right position before the eyes, and the
axes of the eyes and those of the glasses do not correspond, consequently perfect vision is not obtained. (*See
page 107, chap. XII.*) Again, the constant pressure upon
the nasal organ often results in a stoppage of the duct
or canal which leads from the eye to the nose. The
reading glasses, "les binocles" of the French, are objectionable for any but a temporary use, for the reason that
the motions of the head not being in accordance with the
hand which holds the glass, the eyes are very much
tried in attempting to adjust themselves to the varying
positions of the glasses, and this constant strain upon
the eye, to keep itself adjusted, is injurious and should
be avoided.

"But a single eye glass is more injurious still; and
many young men, who, from shortness of sight, or affec-
tation, have thought proper to use a quizzing glass (as

---

* The Fop's eye glass is mounted with Saratoga focus (plane glass), so as to cause no
injury to the eye.

it is termed), have had reason to regret it to the end of their lives. I am acquainted with a gentleman, the sight of whose right eye has been seriously impaired from his having in early life constantly used one of these eye glasses ; and numerous other instances have come to my knowledge. The consequences to perfect vision are serious, for as one eye is made to do more work than the other, an alteration in their relative strength takes place; the result is, that sooner or later when the person resorts to spectacles, he finds that the lens which suits one eye, will not at all suffice for the other.

Watchmakers, and other artists, who work with a magnifier, are very subject to this imperfection of vision, and generally find that they see better with one eye than the other. If, instead of always applying the magnifying glass to one eye, they were to use the other eye in turn, a habit which might be easily acquired in early life, although with difficulty afterwards, they would preserve the power of their eyes more equally, and the perfection of vision longer; for, by using the eyes alternately, rest, and an opportunity of recovering from the fatigue produced by the exertion of looking through the magnifier, would be afforded to each. In like manner, those who indulge in microscopical or astronomical pursuits, should learn to use either eye indifferently, instead of always trusting to one, although we almost instinctively apply the right eye to a telescope or microscope."— *Cooper.*

Many opticians from constantly looking through microscopes, telescopes, &c., with the right eye, thereby render the focus of that eye much longer than that of the other.

Nose glasses are made in thousands of different shapes

and styles, the following cuts representing the principal patterns now in use:

Fig. 25.

Round eye glasses made of horn, shell, rubber, steel, or gold. The rubber is generally preferred, because it is the lighter and more durable; the objection to the horn and shell being that they crack very easily, and to the steel, that rust will soon destroy them. Oval frames are made, of the same style.

Fig. 26.

Fig. 26 represents an oval pattern, with a different style of nose piece.

Fig. 27.

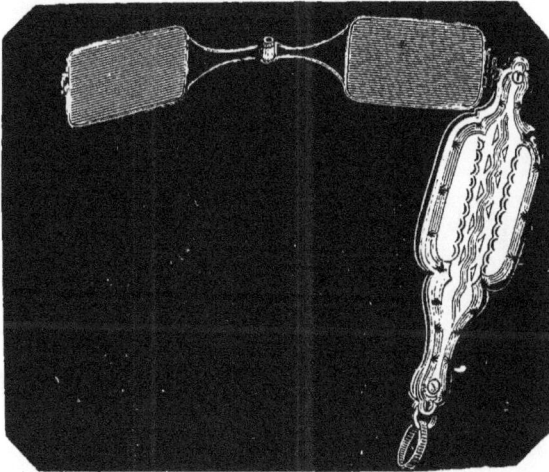

The Lorgnette (Fig. 27), is made on the principle of "les binocles." It is very useful to ladies for temporary use in shopping, &c., or, for myopes, who desire to use glasses only temporarily. The lenses are well protected from scratches, and the lorgnette is an ornament when suspended by a chain, ribbon, &c. They are made of German silver, shell, silver, gold, &c., and are made so that the glasses will fold together into the handle; they open with a spring.

In many cases of cataract, hypermetropia, &c., reading glasses or magnifiers are used to assist vision—the most common kind is the square reader; the lens is

mounted in a frame of brass or German silver, with a
wooden handle, as represented in

FIG. 28.

But the folding Loupe, made of horn or rubber, is often

FIG. 29.

preferred on account of its being more easily carried in
the pocket than the square reader, and without danger
of the lens becoming scratched; but it is not so easily
used, since with the other a line of a column can be read
across without moving the lens.

# RULES,

## FOR THE SELECTION OF SPECTACLES.

Persons purchasing spectacles should always be careful to observe the following rules:—

I. NEVER BUY OF AN INEXPERIENCED PERSON, PEDDLER, JEWELLER, &c., BUT OF AN OPTICIAN.

II. Select the weakest focus with which perfect vision is obtained at the natural distance—say, at 10 or 12 inches from the *eyes*.

III. Remember, that in two or three years from the date of purchase, the lenses, in all cases of presbyopia, will have to be exchanged for those of a greater power.

IV. Select a spectacle frame that is well adapted to the contour of the face, *and that* will bring the center of the lenses directly in front of the center of the pupils.

V. After a perusal of this work, it will not be necessary to caution the reader against the purchase of "Cornea flatteners," " Cornea extenders, &c.," which are only devices to rob the public of their sight and money.

Persons at a distance, or residing in the country, who desire to send to an optician for spectacle-glasses, should always furnish him with the following information ; and if he be a skillful optician he will thereby be enabled to

9

furnish them with the proper spectacles without delay or difficulty :—

1. What is the *nearest* distance at which No. 1 of the Test Types, in this book, can be distinctly seen and read?

2. What is the *farthest* distance at which the same (No. 1) can be read *distinctly*?

3. What is the nearest distance at which No. 11 of these Test Types can be distinctly seen and read?

4. Are letters or objects looked at, in the best range of vision (12 inches), seen distinctly or confusedly, double, or do spots appear to float before the eyes as if a veil was between the eye and the object looked at?

5. Upon keeping both eyes constantly fixed upon an object directly in front of them, can the person at the same time, without moving the eyes in the least, see objects around him and at each side of him?

6. Do you see equally well with each eye? Test each eye *separately*, and furnish the result with each eye to the optician.

7. What is the age of the person?

8. Is the eye myopic, hypermetropic, or presbyopic; and how long has it been so?

9. Are the glasses required for reading, writing and sewing; for seeing objects at a distance; or to protect the eyes from dust, or the irritating action of light?

10. Has any operation ever been performed on the eyes; or have they ever been diseased or treated for disease; what was the operation, or disease; was one or both eyes affected?

11. Have glasses ever been worn before? if they have, for how long a time?

## No. 1.

As they were thus discoursing, they came in sight of thirty or forty wind-mills, which are in that plain; and, as soon as Don Quixote espied them, he said to his squire: "Fortune disposes our affairs better than we ourselves could have desired: look yonder, friend Sancho Panza, where thou mayest discover somewhat more than thirty monstrous giants, whom I intend to encounter and slay; and with their spoils we will begin to enrich ourselves; for it is lawful war, and doing God good service to remove so wicked a generation from off the face of the earth." "What giants?" said Sancho Panza, "Those thou seest yonder," answered his master, "with their long arms; for some are wont to have them almost of the length of two leagues." "Look, sir," answered Sancho, "those which appear yonder are not

## No. 2.

giants, but wind-mills; and what seem to be arms are the sails, which, whirled about by the wind, make the mill-stone go. "It is very evident," answered Don Quixote, "that thou art not versed in the business of adventures: they are giants; and, if thou art afraid, get thee aside and pray, whilst I engage with them in fierce and unequal combat." So saying, he clapped spurs to his steed, notwithstanding the cries his squire sent after him, assuring him that they were certainly wind-mills, and not giants. But he was so fully possessed that they were giants, that he neither heard the outcries of his squire Sancho, nor yet discerned what they were, though he was very near them, but went on crying out aloud: "Fly not ye cowards and

## No. 3.

vile caitiffs; for it is a single knight who assaults you." The wind now rising a little, the great sails began to move: upon which Don Quixote called out: "Although ye should move more arms than the giant Briareus, ye shall pay for it."

Then recommending himself devoutly to his lady Dulcinea, beseeching her to succor him in the present danger, being well covered with his buckler, and setting his lance in the rest, he rushed on as fast as Rozinante could gallop, and attacked the first mill-before him; when, run-

## No. 4.

ning his lance into the sail, the wind whirled it about with so much violence that it broke the lance to shivers, dragging horse and rider after it, and tumbling them over and over on the plain, in very evil plight. Sancho Panza hastened to his assistance, as fast as the ass could carry him; and when he came up to his master, he found him unable to stir, so violent was the blow which he and Rozinante had received in their fall. "God save me!" quoth Sancho, "did not I warn you to have a care of what you did, for that they were nothing but wind-mills?

## No. 5.

And nobody could mistake them, but one that had the like in his head." "Peace, friend Sancho," answered Don Quixote: "for matters of war are, of all others, most subject to continual change. Now I verily believe, and it is most certainly the fact, that the sage Freston, who stole away my chamber and books has metamorphosed these giants into wind-mills, on purpose to deprive me of the glory of vanquishing them, so great is the enmity he bears me! But his wicked arts will finally avail

## No. 6.

but little against the goodness of my sword." "God grant it!" answered Sancho Panza; then helping him to rise, he mounted him again upon his steed, which was almost disjointed.

Conversing upon the late adventure, they followed the road that led to the pass of Lapice; because there, Don Quixote said, they could not fail to meet

### No. 7.

with many and various adventures, as it was much frequented. He was, however, concerned at the loss of his lance ; and, speaking of it to his squire, he said : "I remember to have read that a certain Spanish knight, called Diego Perez de Vargas, having broken his sword in fight, tore off a huge branch or limb from an oak, and performed such wonders with it that day, and dashed out

### No. 8.

the brains of so many Moors, that he was surnamed Machuca ; and, from that day forward, he and his descendents bore the names of Vargas and Machuca. I now speak of this, because from the first oak we meet, I mean to tear a limb at least as good as that ; with which I purpose and resolve to perform such feats that thou shalt

### No. 9.

deem thyself most fortunate in having been thought worthy to behold them, and to be an eye-witness of things which will scarcely be credited." "God's will be done !" quoth Sancho ; "I believe. all just as you say, sir. But, pray set yourself more upright in your saddle : for you seem to me to ride sideling, owing, perhaps,

### No. 10.

to bruises received by your fall." "It is certainly so," said Don Quixote ; "and, if I do not complain of pain, it is because knights-errant are not allowed to complain of any wound whatever, even though their entrails should issue from it." "If so, I have nothing more to

### No. 11.

say," quoth Sancho ; "but God knows I should be glad to hear your worship complain when anything ails you. As for myself, I must complain of the least pain I feel, unless this business of not complaining extend also to the squires of knights-errant."

# INDEX.

—•—

# GLOSSARY.

ACHROMATIC, a combination of lenses so arranged as to correct spherical aberration and prismatic colors.

ALBINO, a Spanish word signifying an individual with white skin and hair; the iris is very pale bordering on red, and the eyes so sensible that they cannot bear the light.

AMAUROSIS, diminution or complete loss of sight without any change of the organization of the eye—generally the patient observes black spots as it were floating before the eye.

CACHECTIC, constitutionally diseased, usually applied to a condition in which the system of nutrition is depraved.

DECUSSATED, crossed, intersected.

DIABETES MELLITUS a disease characterized by a great augmentation and often manifest alteration in the secretion of urine—with excessive thirst, and progressive emaciation—an excessive discharge of saccharine urine.

DIOPTRIC, refracting light; affording a medium for vision.

GLAUCOMA, applied to an opacity of the vitreous humor manifesting itself by a grayish or greenish spot apparently through the pupil—a disease seldom curable.

OPHTHALMOSCOPE, an instrument used in the examination of the interior parts of the eye.

OPTOMETER, an instrument to determine the proper lenses to be worn in myopia and presbyopia.

REFRACTION, deviation from a direct course.

STRABOMETER, an instrument used to readily tell the amount of deviation in a squinting eye.

# ERRATA.

In description of Fig. opposite 1, read "Laminæ," "Lamina."

Page 97, note read chapter VII.

Page 96, between 36–24, read 20 as 30.

# JAMES FOSTER Jr. & CO.
## OPTICIANS

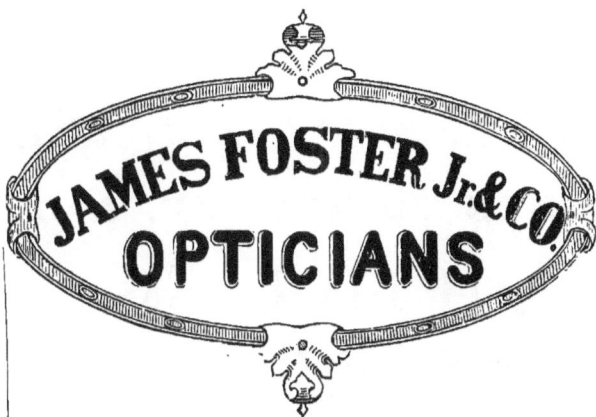

IMPORTERS AND MANUFACTURERS OF

## Optical & Mathematical

## INSTRUMENTS,

## S.W. CORNER
## FIFTH AND RACE
## CINCINNATI. O.

# Dr. J. KING,

# Eye and Ear Surgeon,

## NO. 245 COURT STREET,

### CINCINNATI, O,

Is prepared with all the

## NEW AND IMPROVED INSTRUMENTS,

For examining, and for operating upon

## THE EYE AND THE EAR,

As well as for selecting correct spectacle glasses for persons having long or short sight, astigmatism, &c.

--------

## ARTIFICIAL EYES INSERTED.

# E. SINCERE,

# Optician,

MAIN STREET, UNDER THE NATIONAL HOTEL,

## LOUISVILLE, KY.

---

## Brazilian Pebble Spectacles, Artificial Eyes,

MICROSCOPES,
TELESCOPES,
OPERA GLASSES,
HYDROMETERS,
THERMOMETERS,
BAROMETERS,
SURVEYORS' COOMPASSES
and CHAINS,
STEREOSCOPES and VIEWS,
MAGIC LANTERNS,
—AND—
PHILOSOPHICAL INSTRUMENTS,

## WHOLESALE AND RETAIL.

---

ALL ORDERS PROMPTLY FILLED.

**BENOIT KAHN & CO.**

# Practical Opticians,

535 BROADWAY,

NEAR SPRING ST.,                    NEW YORK.

---

N. B.—Particular attention paid to

## REPAIRING SPECTACLES,

OPERA GLASSES,

*AND ALL ARTICLES IN THIS LINE.*

---

L. SUSSFELD.                    S. LORSCH.                    H. NORDLINGER.

## SUSSFELD, LORSCH & CO.

## Importers & Commission Merchants,

OPTICAL AND FANCY GOODS.

13 MAIDEN LANE, NEW YORK.

**27 RUE PARADIS POISSONIERE, PARIS.**

# F. W. WAGNER & CO.

## IMPORTERS OF

# OPTICAL AND MATHEMATICAL

## INSTRUMENTS,

## 𝕾𝔱𝔢𝔯𝔢𝔬𝔰𝔠𝔬𝔭𝔦𝔠 𝔙𝔦𝔢𝔴𝔰, 𝔉𝔯𝔢𝔫𝔠𝔥 𝔆𝔩𝔬𝔠𝔨𝔰, &𝔠.

## No. 43 MAIDEN LANE,

## NEW YORK.

---

# JAMES PRENTICE,

# OPTICIAN,

## No. 164 BROADWAY,

## NEW YORK.

ESTABLISHED IN - - - 1853.

JAMES W. QUEEN.                                    SAMUEL L. FOX.

# James W. Queen & Co.

# OPTICIANS,

## No. 924 CHESTNUT STREET, PHILADELPHIA, PA.

SPECTACLES,
MATHEMATICAL INSTRUMENTS,
THERMOMETERS,
BAROMETERS,
MAGNETS,
MAGIC LANTERNS,
MICROSCOPES,
AIR-PUMPS,
TAPE MEASURES,
DISSOLVING APPARATUS,
WORKING MODELS OF STEAM ENGINES,
SPY GLASSES,
OPERA GLASSES,
ELECTRICAL MACHINES,
&c., &c., &c., &c., &c.,

## WHOLESALE AND RETAIL.

ILLUSTRATED CATALOGUES SENT GRATIS.

# PAUL ROESLER,

# OPTICIAN,

## NEW HAVEN, CONN.

Keeps constantly on hand and for sale,

## AT WHOLESALE AND RETAIL,

SPECTACLES,
      OPERA GLASSES,
            SPY GLASSES,
                  THERMOMETERS,
BAROMETERS,
      HYDROMETERS,
            DRAWING INSTRUMENTS,
                STEREOSCOPES, &c.

---

☞ Particular attention is called by the subscriber to his

## NEEDLE POINTED DIVIDERS,

for the use of Schools.

## PAUL ROESLER,
### NEW HAVEN, CONN.